Principles of

Pneumatic Architecture

Principles of Pneumatic Architecture

Roger N. Dent
B Arch (Hons)
of Building Design Partnership

The Architectural Press, London

85139 068 4

First published 1971

Made and printed by offset in Great Britain
by William Clowes & Sons, Limited
London, Beccles and Colchester

To Jane

Acknowledgements

In preparing this publication I have depended a great deal on the help and advice of many people and organisations, to whom my sincere gratitude goes as follows.

To Building Design Partnership and the partners who suggested I should venture into the publishing world and who have enthusiastically encouraged me, to William Brook who has spent much time reading the drafts and given me much guidance, to Eric Bassett for his help with chapter 3, to Marjorie Willan for her successful searches for obscure references, to the typists, especially Georgette Bassant, Teris McWilliam, and Linda Hill, to my close friend John C. Pye who spent many hours drawing out the illustrations which greatly clarify the text.

To Professor Rawlings, not only for writing the Foreword but also for his comments and advice on the text.

To Professor Markus, School of Architecture, University of Strathclyde, who as my tutor at the University of Liverpool fostered my initial enthusiasm for pneumatics and has continued to help and advise me.

To the Architectural Press, London and Godfrey Golzen for much help and guidance.

Besides these my thanks are due to the individuals and organisations listed below, some of whom have provided photographs, and others who will remember correspondence. . . .

Sir Vivian Fuchs, Walter Bird, Arthur Quarmby, Cedric Price, Frank Newby, Dr. R. H. Suan, Norman Foster, E. Fritzsche, E. Macher, Manfred Schiedhelm, François Baschet, Yutaka Murata, Victor Lundy, Carl Koch and Associates, Brigham Eldredge Limon Hussey, Davis Brody and Associates.

Scott Polar Research Institute Cambridge, National Aeronautics and Space Administration, National Research Council of Canada, Ministry of Technology Research and Development Establishment Cardington, Military Vehicles and Engineering Establishment Christchurch, American Polar Society.

Birdair Structures Inc., Bayer Chemicals Ltd., Gourock Ropework Co. Ltd., Computer Technology Ltd., Lea Bridge Industries Ltd.,

Don Cameron Balloons, Du Pont Company, Cross and Ticher Ltd., Fried. Krupp Universalbau, British Hovercraft Corporation Ltd., Goods and Chattels Ltd., Robert Wilson and Sons Ltd., Frankenstein Group Ltd., M.L. Aviation Company Ltd., Taiyo Kogyo Co. Ltd., Kurashiki Rayon Co. Ltd., Barracudaverken, L. Stromeyer and Co., Hovercraft Development Ltd., Firestone Coated Fabrics Company, Goodyear International Corporation, Garrett Corporation, R.F.D.-G.Q. Ltd., Pneumatic Tent Co. Ltd.

In addition I am grateful to the following for permission to quote from their publications, International Association for Shell Structures, Professor Otto, Professor Oden, Nickolaus Laing, Cassell and Co. Ltd., McGraw Hill Inc., Macmillan Company.

Foreword

Pneumatic structures are a source of great interest to architects, engineers and in fact to the community at large. We support our motor cars on them, propel our sailing boats by them, and rely on them to save our lives if we have to parachute from an aeroplane or bob up and down on the waves in an inflatable life raft. We encounter them early in life in the form of rubber balloons and probably learned to swim with an inflatable ring which provided buoyancy.

Nowadays we can enter a large 'airhouse' and play indoor games in complete comfort, sheltered from the inclement weather outside; alternatively, we can use it for storage purposes or for the protection of civil engineering construction work. However, as will be seen in this book, the now familiar 'airhouse' is just one, perhaps very early, stage in the development of pneumatics for architectural purposes. Mr. Dent has, of course, discussed the 'airhouse' in some detail, but he has also delved deeply into the architectural implications of pneumatics, in what I believe to be the first book devoted to this subject. The book, however is not of interest only to architects, but should stimulate the thoughts of all concerned with the building industry.

In studying the development of pneumatic structures it quickly becomes obvious that they have evolved quite independently of conventional structures and the present state of the art is one which ranges from high quality forms such as the U.S.A. pavilion at EXPO '70 in Tokyo, down to cheap goods, of doubtful quality and limited durability. The image which pneumatics have is, regrettably, tarnished by the prevalence of the latter, but in the same way that we do not condemn a well-designed bridge or building because we fell off a poorly designed chair, we should not be prejudiced by any poor quality pneumatics which may appear from time to time, but be prepared to look afresh when we evaluate the new developments in this field. And many new developments there will be!

In this book Mr. Dent has given the reader an insight into the many new concepts which are emerging, and foreshadows some of these new structural forms. After reviewing the developments which have taken place, he gives consideration to current practices, requirements and limitations of materials, inflation equipment, access, safety and other aspects of current interest, and devotes the latter chapters to applications of pneumatic structures which have attracted attention in recent years. The sections on polar shelters, pneumatic formwork and the recent EXPO '70 structures are particularly interesting.

I have found this book to be informative and fascinating yet light and interesting to read. I do hope you find it as absorbing as I did.

Barry Rawlings, Professor of Civil and Structural Engineering, University of Sheffield, October 1970.

Contents

Terminology

A Pneumatic Structure
Any structure which is supported or motivated by the action of pressure differentials, created with air or gases.

An Air Supported Structure
A single membrane structure which is supported by an air pressure slightly above that of the atmosphere.

An Air Inflated Structure
A structure in which the air is contained by a membrane to form inflated structural elements, such as columns, beams, arches and walls.

An Air Controlled Structure
A structure whose position or movement is controlled by the action of air pressure differentials.

Notations

P — *Total pressure differential across the membrane at a point.*
P_a — *Average pressure differential across the whole of the membrane.*
P_i — *Internal pressure acting on membrane.*
P_x — *Total pressure acting on membrane due to external loads only.*
P_w — *Pressure acting on membrane due to wind loads.*
P_s — *Pressure acting on membrane due to snow loads.*
P_g — *Pressure acting on membrane due to dead loads.*
G — *Dead loads.*
S — *Snow loads.*
Q — *Dynamic impact pressure of wind.*
C — *Aerodynamic pressure coefficient for wind loading.*
R — *Radius of curvature of membrane.*
T — *Membrane stress.*
T_l — *Longitudinal membrane stress.*
T_c — *Circumferential membrane stress.*
T_m — *Maximum limit stress of membrane material.*
T_p — *Usable design stress of membrane material.*
Y — *Total safety factor.*
Y_1 — *Safety factor accounting for material strength loss.*
Y_2 — *Safety factor accounting for building function.*
v — *Wind speed.*
k *is used as a constant.*

NOTE

Throughout this book the S.I. Units of the Metric System are used. For pressure this implies the use of the rather unfamiliar unit N/m^2 (Newton/square metre), which is equivalent to 9·807 kgf/m^2 (kilogramme force/square metre).

1. An Introduction to Pneumatics

Technological development nowadays is so rapid that soon the innovations of today will be obsolete tomorrow. This era of change is naturally reflected in social trends. People, particularly the young, are no longer satisfied with the same old environment day in and day out. They demand change and variety, and this is manifest in two ways: firstly the greater turnover in material belongings, such as cars, furniture and clothes, and secondly the increased movement, not only from one occupation to another but also between physical environments. The latter has been encouraged by the advances that have been made in communications. How then should these social, technical and economic changes be reflected in architectural design? A new versatile architecture must surely emerge which can accommodate these trends, perhaps a portable instant architecture or even a short-term throw-away architecture. Traditional architecture, evolving as it does from rigid structural forms which dictate the environmental conditions within them, could hardly be adapted to suit these requirements. However, a new architecture has been born, that of pneumatics, which is infinitely more flexible in its options. It can be erected or dismantled quickly, is light, portable and materially inexpensive. It therefore offers a possible solution to a wide range of problems, both of social and commercial kind. For instance, pneumatic construction can be used to overcome temporary shortages of warehousing space. It can also be used to provide shelter for the homeless in times of natural or man-made disaster, and in these early days of space exploration, it has even been suggested for lunar shelters. But of more importance than these applications demonstrate, is the fact that pneumatic construction points the way to an architectural revolution. To correct the environmental deficiencies of rigid traditional structural envelopes, energy must be supplied to heat and ventilate them, bringing them up to the comfort standards that are determined by the building's function; the amount of this applied energy depends on the insulation characteristics of the structural envelope. Advances in technology have increased the effectiveness of these characteristics, but at the same time environmental engineering has become a much more sophisticated science. It is now possible to create a

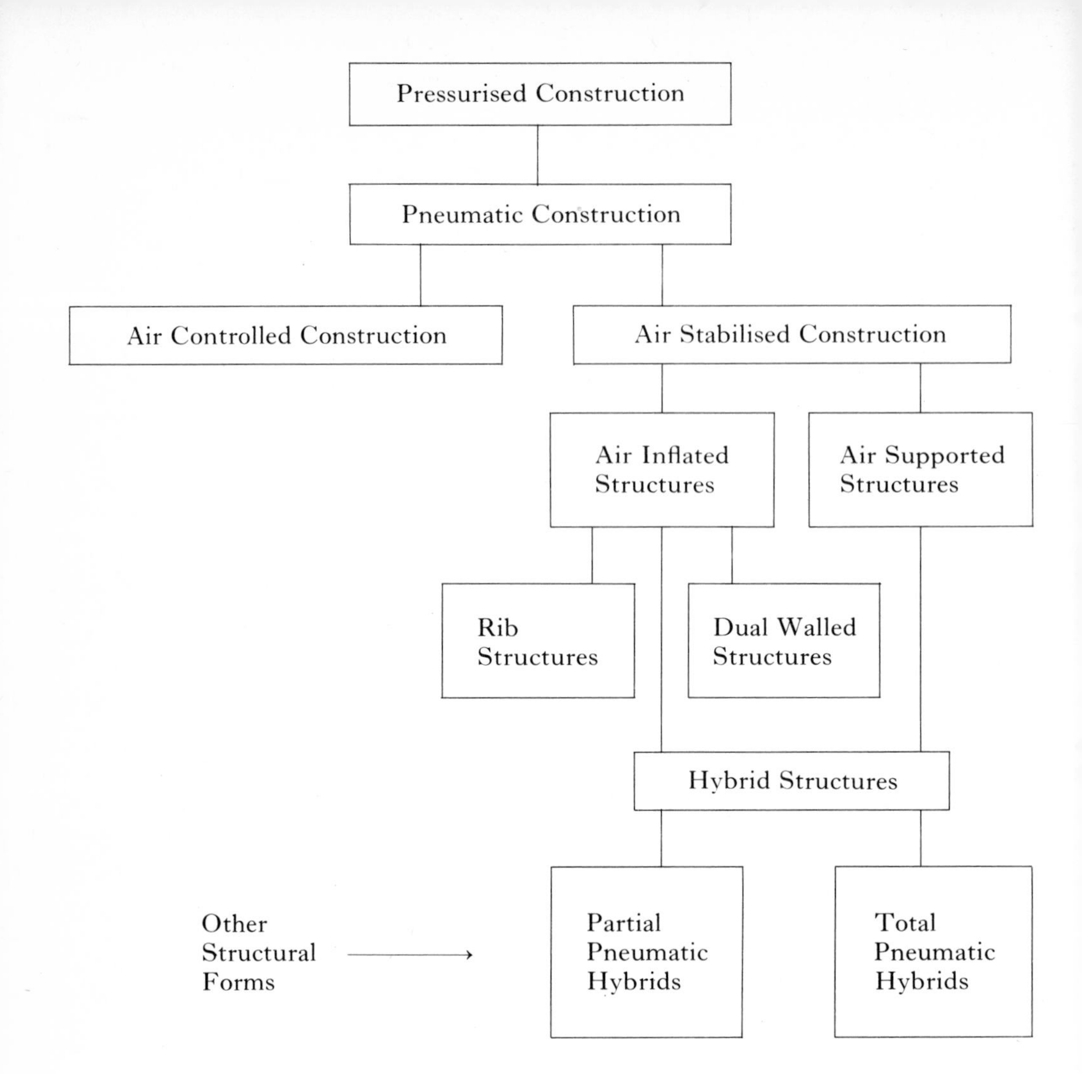

Fig. 1

fully conditioned environment without a structure of the kind usually associated with architecture; all that is required is a bag to contain this manufactured environment. This is the architectural essence of a pneumatic structure; a membrane or bag stabilised by small pressure differentials created by the application of environmental energy. Such a concept, as pointed out by Professor Reyner Banham,[1]* is a complete reversal of architectural thinking; on the one hand there is traditional or conventional architecture in which structure determines environment, and on the other hand there is pneumatic architecture in which the application of environmental energy produces structural stability.

Until recently pneumatics have been something of an architects' and designers' toy—an intriguing novelty at Expos' and such like.

* Throughout the book, full details of references are given at the end of chapters.

It is now becoming plain, however, that their potential is a great deal more serious than this. However, despite growing interest in the subject, on the part of both architects and engineers, which is reflected in numerous magazine articles, very few books have been written about it and none of these have dealt specifically with its architectural implications. It is hoped, therefore, that this book will fill a gap. Its aims are not only to describe the principles of pneumatic construction and the present state of pneumatic technology, but also to make a wider circle of readers aware of the real potential of this exciting new structural medium.

DEFINITIONS

As with many technologies in their infancy, the terminology of pneumatics has not been clearly defined, although certain terms appear to have been accepted throughout the world. The words 'pneumatics', 'blow-ups', 'inflatables', 'airdomes', 'airhouses', and many others are tossed around rather nonchalantly to describe in one case the whole field of this technology and in another just one particular aspect. To define it accurately, however, it should be known collectively as *pressurised construction*,[2] a term which implies the control and stabilisation of all kinds of structures by means of pressure differentials achieved by the uniform loading actions of air, gases, liquids, or even granular solids. Fig. 1 illustrates the relationship between the general category of pressurised construction, and those sub-categories of it which, under the general heading of *pneumatic constructions* refer to structures acted on by air or gases and relate particularly to architecture and building. The reader may find it helpful to refer back to fig. 1 in the ensuing paragraphs which briefly explain the principles on which each sub-category of pneumatic construction operates.

Air Controlled Construction

Any structure whose position or movement is controlled by air pressure differentials can be termed 'an air controlled structure'. As yet these structures are not generally associated with architecture, although pneumatic tubes have long been used to transport messages and cash in big department stores. More familiar applications are pneumatic drills, air braking systems, and ground effects machines, such as hovercraft and hoverpads. These latter ground effects machines may be of particular significance in the architecture of the future, and this possibility is examined more closely in later chapters.

Air Stabilised Construction

As yet, only the air stabilised structure has been adopted for building construction. The essence of pneumatically stabilised construction, whether it be stabilised by air or gases, is a thin flexible membrane, which is supported solely by pressure differentials. These differences

Fig. 2

in pressure induce tensile stresses into the membrane, and enable it to support gravitational wind loads as a relaxation of these tensile forces. Consequently the pneumatic structure is a pure tensile structure, in which the membrane material is utilised with great structural efficiency. An analogy of this phenomenon is the water hosepipe. This demonstrates the pneumatic principle very simply. When empty it is limp and possesses very little stiffness, but once filled with water it becomes more rigid. The water is causing a pressure differential across the walls of the hosepipe, which are so pretensioned as to resist bending.

Although this construction principle is very straightforward and easily grasped, its very simplicity is misleading. A full understanding of its pneumatic behaviour is most complex and has yet to be achieved with any accuracy. Basically, however, there are two conditions that govern this form of construction.[3] Firstly, the pressure differential inducing the membrane tension must be high enough under the action of all loading conditions to prevent compressive membrane stresses; such compressive stresses appear as folds in the membrane. In other words, complete stability is achieved when all parts of the membrane are under tension. Secondly, the membrane stresses at any one point under all loading situations must be less than the permissible stress of the material. These conditions can be represented by the following expressions:

$$T_{min} \text{ `} P_x + P_i \text{'} \geqslant 0 \qquad (1.1)$$

$$T_{max} \text{ `} P_x + P_i \text{'} \leqslant T_p \qquad (1.2)$$

If the pretension stress in the membrane is exceeded, producing compressive membrane stresses, the membrane merely deforms so that the stresses are redistributed and the above condition (1.1) is again satisfied. If the membrane is prevented from deforming, these stresses cannot be redistributed, and therefore compression folds occur in the membrane. Thus the pneumatic structure is dynamic and alive, as against the statics of more traditional structural concepts. Here lies the complication in the design of pneumatic structures; this distortion must be understood, yet controlled in a manner that keeps stress concentrations within allowable limits.

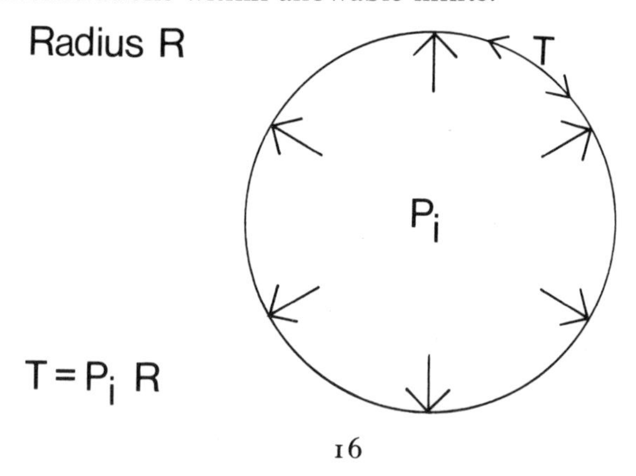

Fig. 2

Arising from the principles of air stabilised construction, two types of structure emerge: the air inflated structure and the air supported structure.

Air Stabilised Construction: The Air Inflated Structure

In air inflated construction, air is contained within a membrane to form inflated structural elements, such as columns, beams, walls and arches, which themselves resist the external loadings in much the same way as the structural elements of more conventional structures. *Fig. 3* The structural capabilities of these inflated elements are dependent on four things, the volume of air contained within the element by its membrane, the excess pressure differential exerted on the membrane by the air, the characteristics of the membrane material, and the structural form of the element. With larger volumes and higher pressures greater spans can be achieved, but these necessitate the use of stronger membrane materials. The main characteristic affecting the structural performance of the element, besides tensile strength, is the modulus of elasticity of the materials; the higher this is, the greater is the rigidity of the structural element. If the air could be sealed within the membrane, and no air leakages occurred, then air replenishment would not be necessary to maintain a constant pressure. However because of the slight air porosity of materials generally used in this type of construction, particularly at joints, air replenishment is generally essential. In addition to this, temperature variations cause expansion and contraction of the air and this effects a variation of air pressure. With the higher pressure structures this effect is not so pronounced, but on the other hand with very low pressure structures it is very marked and can cause collapse of the structure due either to reduced air pressures or overtensioning of the membrane material by increased air pressures. Consequently periodic air replenishment is needed, and in cases where large air volumes are involved, a continuous air supply is sometimes essential.

There are two main types of air inflated structure, inflated rib structures and inflated dual walled structures. The former is made up of a framework of pressurised tubes which supports a weatherproof membrane in tension. This membrane can add considerably to the stability of the structure. The small volumes of air contained within these tubes make these structures more suitable for small span construction.

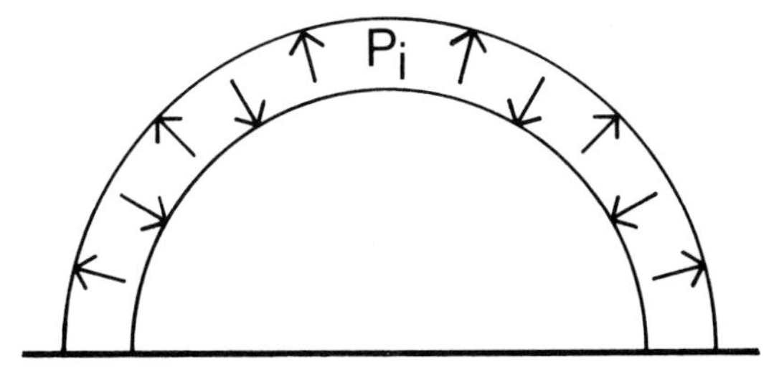

Fig. 3

Plate 1 A PTC air inflatable rib shelter

Inflated dual walled structures, as their name implies, consist of two membrane walls between which the air is contained. These walls are held together by drop threads and diaphragm configurations. The larger volumes of air associated with this form of construction give it a greater spanning potential than air inflated rib construction.

Plate 2

Erection times for air inflated structures vary considerably and are dependent on the volumes of air contained within the membrane, the pressure of this air and the capacity of the inflation equipment. Small structures like air inflatable tents can be inflated within a couple of minutes, an instant structure in fact, whereas the larger structures can take thirty minutes or even longer.

Air Stabilised Construction: The Air Supported Structure

Air inflated construction, apart from the fact that it depends on air pressure differentials for stability, cannot really claim to be far removed from conventional structural systems. Air supported construction, on the other hand, is certainly a completely new structural form which in some cases has already shown signs of producing a complete contradiction of traditional architecture.

Fig. 4

An air supported structure consists of a single structural membrane which is supported by a small air pressure differential. This means that the internal building volume is at a pressure slightly above atmospheric, usually between 15 and 25 mm of water pressure, and consequently access in and out of the building is accomplished across a pressure differential. There will thus be a continuous leakage of air from the building interior, which must be replenished by an un-

Plate 2 A travelling exhibition pavilion for the United States Atomic Energy Commission, of dual walled inflated construction, and manufactured by Birdair

interrupted air supply so that the pressure differential is maintained at all times. Unlike conventional structures which exert a positive loading on the ground, the pressure differential across the membrane of an air supported structure causes up-lift forces, and these must be resisted by firmly anchoring the air supported structure to the ground. Here then are the three unique structural design problems presented by the air supported structure; firstly the need to maintain the pressure differential across the membrane with a constant air supply; secondly the need to minimise air leakages; and thirdly the need to counteract the up-lift forces with some means of anchorage. As with the air inflated structure, the strength of the air supported structure is dependent on the contained air volume, the internal air pressure, the structural form and the membrane material characteristics, the former two being the most influential; the greater these two variables are, then the more rigid the structure is. Thus the larger volume, enclosed by the membrane of the air supported structure, requires only very small pressure differentials, 15 to 25 mm of water pressure,

Plate 3

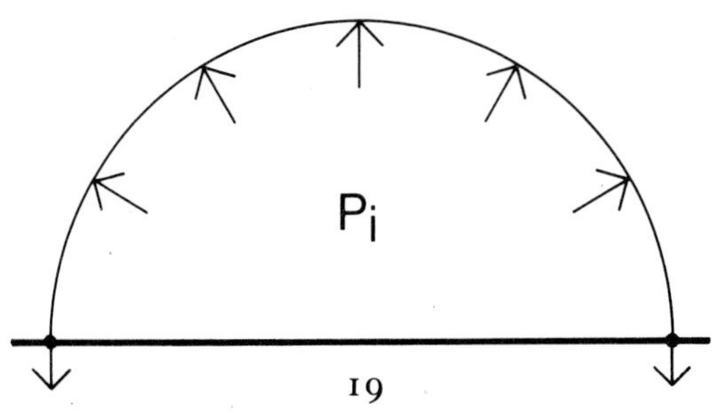

Fig. 4

Plate 3 A typical air supported sportshall, fabricated by Barracuda

for stability. With air inflated structures smaller volumes are contained by the membrane material, and so higher pressure differentials are required. From this it can be seen that air supported structures possess the greater spanning potential. However, against this great advantage must be set the three unique design problems of air supported construction, which have already been mentioned.

Behaviour of Air Supported Structures

Unlike the materials used in conventional structures, the membrane of air supported structures is not directly resisting the applied external loads, but is containing the air so that the action of air pressure is forming a stable structure against external loadings. If the external loads were uniform, then an equal internal pressure differential would support these loads directly, leaving the membrane as just a separating medium completely free of any tensile stresses. Since, in this instance, no stresses are transferred into the membrane material, no anchorage of the structure is necessary. This means that, unlike other structural

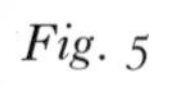

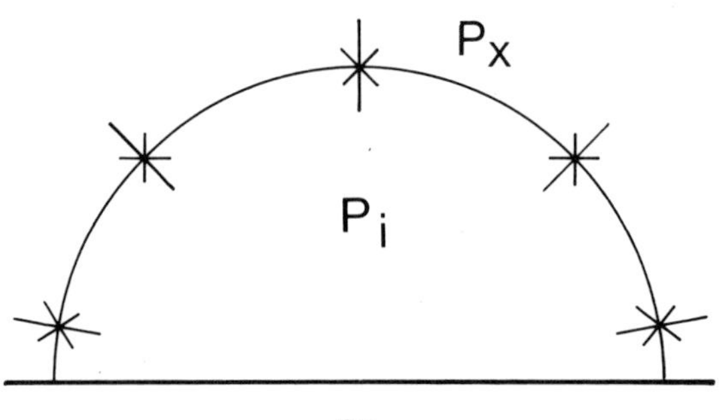

Fig. 5

concepts, there is no theoretical limit to the span of such an air supported structure, as determined by the characteristics of the membrane material. However, in practice, the external surface loadings are never uniform, and so the pressure differential must be high enough to prevent compressive stresses in the membrane. Even taking into account a one-sided external loading, no one has yet been able to set a limiting span for air supported construction, but it is certainly far greater than any other known structural spanning device.

Air supported structures, in many ways, can be likened to a human being. One has only to occupy such a structure for a short while, to be aware of its response to minute variations in climatic parameters and loading conditions. Strike it and it visibly deflects with 'pain'; no static structure is this. It is alive and, like the human body, needs to retain a specific inherent energy level, or pressure level; this it does by means of a continuous air supply. The inflation equipment is the 'heart' of the air supported structure, and if this malfunctions, stopping the air supply, the structure will deflate and slowly 'die'. But like the human body it is very resilient and can be 'resuscitated'. When the inflation equipment has been repaired, the structure will reinflate, provided that the structural membrane has not been seriously damaged during the collapse. Can any other structure be re-erected so quickly following collapse?

Basic Elements of Air Supported Structures

As far as basic essentials are concerned, the air supported structure is made up of four elements, the structural membrane, the means of supporting this membrane, the means of anchoring it to the ground, and the means of access in and out of the building structure. Generally the membrane is fabricated using a coated synthetic fabric, like p.v.c. coated nylon, and this is supported by a pressure differential maintained by a constant supply of air provided by simple low pressure fans. Although there are many possible anchorage methods, the membrane is generally clamped firmly to a concrete foundation. For ease of access against the pressure differential, air locks are necessary. These are the basic structural elements of air supported construction and they will be discussed in much greater detail in a later chapter.

Hybrid Structures

Because of the limitations of both air supported and air inflated construction, another structural form has emerged, the hybrid structure. The term 'hybrid' covers a very wide range of structural types, although two basic forms can be clearly distinguished. The first has logically developed from these limitations, but has capitalised on the benefits of each of these two constructional forms by integrating them both into one structural system. The spanning potential of air supported construction is achieved, but with the added insulation of dual wall construction. Since structural stability is achieved by two

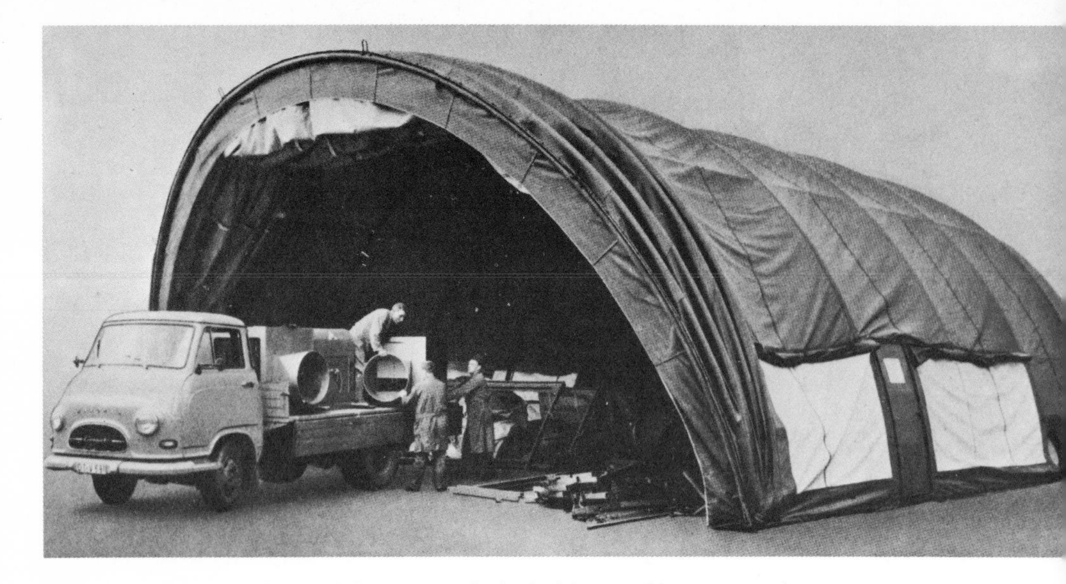

Plate 4 A partial pneumatic hybrid; an air supported structure manufactured by Krupp with a secondary tubular frame construction

different methods, safety against collapse is increased. As yet there are very few of these sophisticated structures, but amongst them is one of the most notable examples of pneumatic architecture. This building, Lundy's exhibition pavilion for the United States Atomic Energy Commission is discussed at length in the following chapter.

Whereas this first group might be termed 'total pneumatic hybrids', the second group would be described as 'partial pneumatic hybrids'. These structures combine pneumatic construction with other more conventional forms of construction, and obviously the variation and range of combinations is immense. The semi-rigid airships of the early twentieth century are but one example of this type of hybrid construction. The local shape of the airship was maintained by the internal pressurisation, whilst a very light keel provided overall rigidity of the structure. Although similar constructions have been used with air supported structures, the initial idea was not to produce an integrated structure like the airship, but to supplement the air supported structure with a light metal structure to overcome some of the problems posed by air supported design. These problems, namely those of providing a continuous air supply and access against the pressure differential, are surmounted by designing a membrane structure which under very light loading is supported and tensioned by a lightweight metal frame. Under heavy loads of wind and snow the interior is pressurised to support this increased loading. This allows easy and large access provision during windless periods. However, although the air supply is not in continuous use, its reliability must be ensured, since automatic activation is necessary when the loading of the structure reaches the critical value at which the metal frame structure can no longer support the membrane on its own. With air

Plate 4

inflated structures other structural techniques, such as tension construction and frame construction, are integrated to increase both the spanning potential and rigidity of air inflated construction. Almost certainly, it is through hybrid construction that the truly sophisticated pneumatic architecture of the future will emerge.

REFERENCES

1. Reyner Banham, *The Architecture of the Well-tempered Environment*, p. 274.
2. F. Otto, *Zugbeanspruchte Konstruktionen*, Band 1, p. 10, uses the term 'pneumatics' to define all types of pressurised construction.
3. R. Trostel, 'On the Analysis of Membranes', *Proceedings of the 1st International Colloquium on Pneumatic Structures*, University of Stuttgart, 1967, p. 84.

2. The Development of Pneumatics

Beginnings in Nature

Although it is only recently that pneumatic construction and the use of air as a supporting medium has become a part of architectural language, pneumatic technology is by no means a newly found science. Man has used pneumatic techniques from very early times, no doubt from his observation of natural pneumatic structural forms in both plant and animal life. In many of nature's structures, fluids under pressure are contained within thin flexible membranes. In man's body, blood is pressurised and enclosed within thin and very flexible skin tissues, which remain taut under the action of this blood pressure. Air bubbles formed in liquids are undoubtedly nature's most relevant precedent in the study of pneumatic building construction and these will be discussed in the following chapter. Although man soon adapted the animal skin for water storage, another pneumatic structure, the sail, was probably his first attempt to utilise air pressure differentials. Having taken to the water, he discovered that, with the sail, he could harness the wind's energy to provide him with a means of propulsion. Aerodynamic differences in pressure, due to the wind, cause inflation of the sail into a billowing form.

Balloons and Airships

In comparison the balloon, a pneumatic more akin to air buildings, is a more recent development. Even so, as early as the thirteenth century, Roger Bacon put forward the idea that a huge globe of thin metal would rise to the heavens when it was filled with the very thin air of the upper atmosphere.[1] Although this idea conveniently omits to state how the thin air could be obtained, the principle of the basic balloon was stirring in man's mind. During the seventeenth century, similar ideas for a balloon, based on the use of a substance lighter than air or on the use of a vacuum, were put forward by Cyrano de Bergerac and Father Francesco Lana.[2] Leading scientists experimented extensively with soap bubbles and paper bags, which ascended when filled with hydrogen. Judging by the decoration of classical earthenware, which often shows children blowing soap

Plate 5 The balloon, a rare sight these days

bubbles through a pipe, the ancient civilisation of Egypt, Greece and Rome were no doubt fully aware of this phenomenon. However, it was not until the eighteenth century that the balloon became a reality. In 1783 the Montgolfier brothers inflated, with hot air, a 10 m diameter sphere made of paper and linen, and they observed this sphere rise to a considerable height before it descended. At the same time as these experiments of the Montgolfiers, Jean Baptiste Meusnier was suggesting a design for a dirigible non-rigid airship, which was even more revolutionary.[3] His designs were for a cigar shaped structure with an inner bag, containing hydrogen as the lifting agent, surrounded by an outer envelope containing air at a higher pressure than that of the atmosphere. The outer envelope was thought necessary in order to give the airship enough rigidity to withstand the severe dynamic loads imposed by wind, air pockets and turbulence. These were the beginnings of pneumatic air transportation, and were considerably developed until the twentieth century, when the advent of

Plate 5

Plate 6 Airships are still manufactured by the Goodyear International Corporation

the aeroplane made them obsolete for most practical purposes. Although they have largely disappeared from the sky, except as use for advertisement gimmickry, scientific observation and as a leisure sport, they are of considerable significance in pneumatic building. They are unfortunately remembered vividly for the many disasters that befell them. These disasters, most of which occurred with rigid airships, were caused by firstly the use of highly explosive hydrogen gas and secondly because of the failure of designers to fully comprehend their complex structural behaviour. However by the time the reign of the airship was over many of these structural problems had been solved. Initial development of pneumatic buildings gained much from these techniques which were used in balloon and airship design, and it is interesting to note that Walter Bird, the great pioneer of pneumatic buildings, moved into this field of engineering from that of aeronautics. However, some of the ingenious solutions that airship designers used in overcoming such problems as weight, stiffening, waterproofing and structural shaping have yet to be fully exploited in the design of pneumatic buildings.

Plate 6

The Pneumatic Tyre

In one field pneumatic construction has without doubt long established itself as the best solution to a particular problem, that of providing motor vehicles with a smoother ride. The pneumatic tyre, although patented as early as 1845 by Robert William Thompson, was not universally adopted for cycles and motor vehicles until the beginning of the twentieth century. Its main advantages over the solid tyre are twofold, firstly, its superior ability to absorb road shocks through considerably greater deformation, and secondly its unrivalled handling characteristics due to the fact that a greater surface area of tyre is in contact with the road surface. In this instance, pneumatic construction has been proven and accepted as structurally sound, but like many other pneumatic forms it is now taken for granted and not really associated with pneumatic buildings which use similar structural principles.

Lanchester; the Architectural Beginnings

The first known architectural attempt to apply the balloon principle to earthbound structures was by the English engineer, Frederick William Lanchester. In his patent of 1917 for a field hospital, the basic principles of air supported construction for buildings were realised.[4]

> 'The present invention has for its object to provide a means of constructing and erecting a tent of large size without the use of poles or supports of any kind.
>
> The present invention consists in brief in a construction of tent in which balloon fabric or other material of low air permeability is employed and maintained in the erected state by air pressure and in which ingress and egress is provided for by one or more air locks. . . .
>
> *Fig. 6*
>
> In one mode of carrying the present invention into effect a rectangular sheet of balloon fabric suitably reinforced by bands, ropes or nets, is pegged to the ground along two of its parallel edges, a flap being left beyond the point of attachment which is turned under. The ends of the said rectangular sheet have stitched to them extensions cut after the manner of a spherical balloon to form quadrant segments of spherical or approximately spherical form. These extensions likewise have a marginal flap which is turned under.
>
> *Fig. 7*
>
> The whole of the above having been securely staked and if necessary loaded by ballast the interior is inflated by moderate air pressure by a centrifugal fan, and the whole so inflated forms a tent of segmental form terminated by domelike ends. The marginal flap initially turned under in laying out the envelope now forms an air seal in contact with the ground, and where necessary is loaded by sandbags in order to maintain it in close

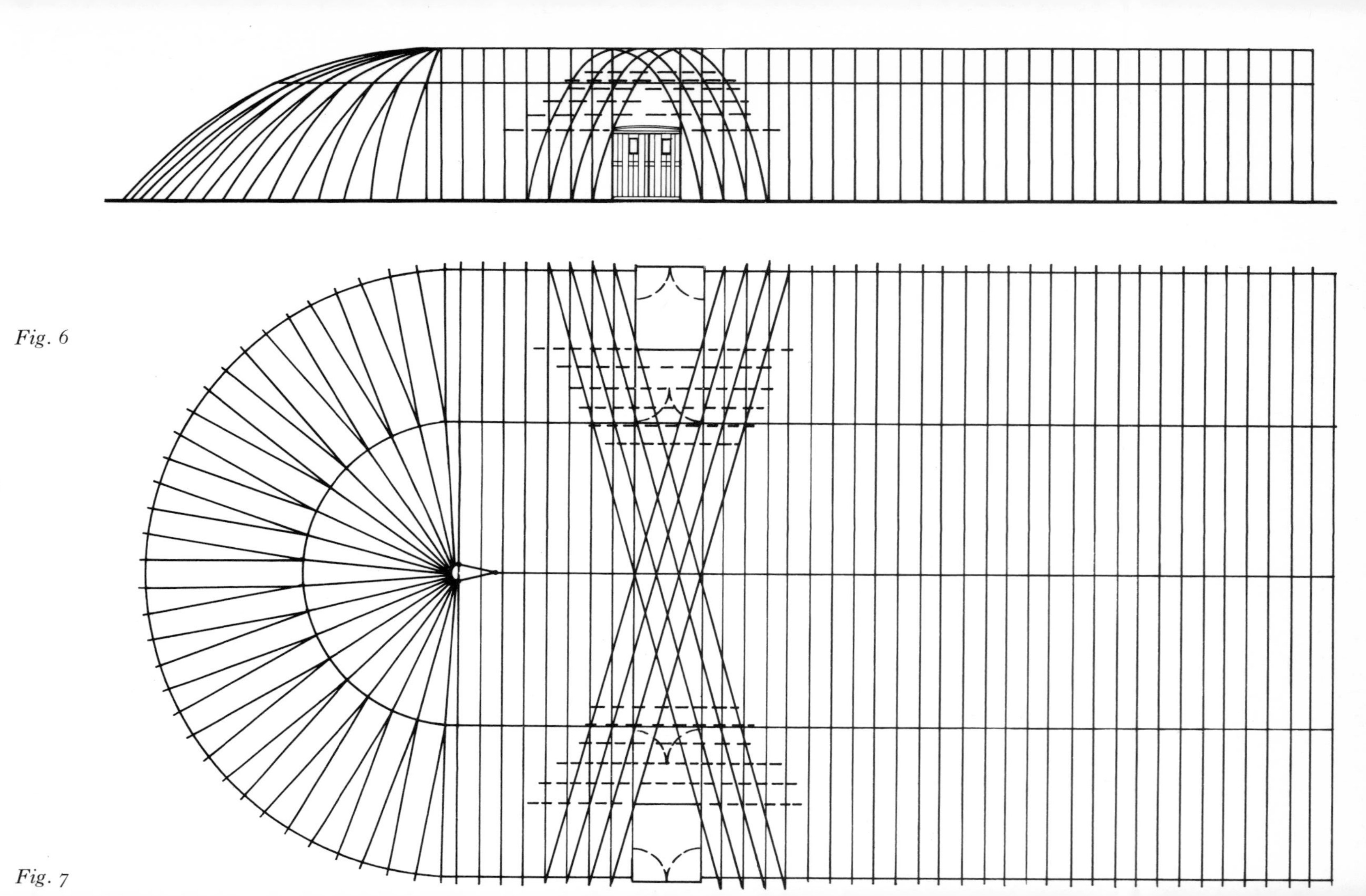

Fig. 6

Fig. 7

contact and minimise air leakage. One or more doors in the form of an air lock, constructed as hereinafter described, are arranged at suitable points according to the use to which the tent is required to be put. The technique of construction in such matter as roping, netting, reinforcing, etc., may closely follow similar well-known methods at present practised in connection with dirigible balloons.'

Lanchester also suggests that a pressure differential of between 25 mm and 75 mm of water pressure is all that is required to maintain structural stability, the precise magnitude of this pressure differential being related to external loads, such as wind and snow.

This patent, clearly derived from balloon and airship construction, is remarkable on two accounts: firstly, he appears to be fully aware of all the basic implications of buildings supported by air, and secondly, although his patent concerns a field hospital, he mentions the potential of air supported buildings for huge spans such as those encountered in aircraft hangars and sports arenae. Later, in 1938, he looked more closely into the potential of pneumatics for large span construction.[5] With his brother Henry V. Lanchester, F.R.I.B.A., he prepared design proposals for an exhibition building, 330 m in diameter. In this design, nets and cables not only reinforce the structural membrane, but also influence the structural form. Unfortunately Lanchester was never able to see the realisation of any of these designs. It was 30 years later before the first air supported structure was erected. Many people feel that this long lapse of time is due to the lack of suitable materials and techniques to accomplish these proposals. However, this explanation is rather naïve, since Lanchester's proposals only made use of the technology available to him at that time. Lanchester himself put forward a much more plausible reason.[6] When he was first connected with industry, before the turn of the

Fig. 8

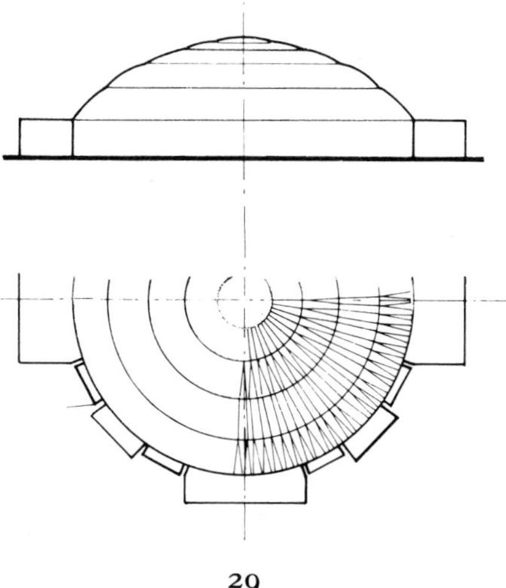

Fig. 8

Plate 7 The PTC air inflatable tent, a common feature on camping sites

century, men were full of vigour and enterprise, and many people were anxious to explore new ideas. Soon after his patent, came the Armistice of the First World War, which in turn was followed by a deep industrial slump. Such was the magnitude of this slump, that few would risk experimenting with new ideas. His ideas were so opposed to the static and heavyweight concepts of construction, that other designers of the day could not accept them. Even today the use of air as a supporting element is still regarded as rather utopian. Further amplifying these doubts of his contemporaries, were the many disasters that befell both balloons and airships. Despite Lanchester's failure to realise his proposals, surely he can justifiably be called the originator of pneumatic building construction.

Pneumatic Camping Equipment

Probably the first pneumatic structure to hint at the possible future role of pneumatics for building construction was the Stromeyer inflatable camping tent, consisting of a waterproof membrane stretched between a pair of intersecting air inflated ribs. Such tents, now a very common feature on camping sites, first appeared before the Second World War, exemplifying a very feasible form of construction for portable buildings. Material bulk is kept to a minimum by deployment of material purely in tension, the most efficient structural use of material possible. In addition to this, a minimal

Plate 7

energy expenditure is required for erection as well as dismantling of the structure for future re-use. These two characteristics make pneumatics one of the most portable constructional concepts known to man, and undoubtedly account for the popularity of inflatable camping equipment, such as tents, beds and chairs.

Stevens' Proposals for an Air Supported Roof

During the Second World War, Herbert H. Stevens, an American engineer, tried to promote Lanchester's ideas with a design proposal for a pneumatic building to house the manufacture of aircraft.[7] The design consisted of a thin steel membrane, held up by air supplied from sixteen fans and anchored to a concrete ring beam. The roof could be constructed on the ground, out of strips of 18 gauge sheet steel, welded together electrically. This sheeting could then be welded to steel plates attached to reinforcing rods in the concrete ring beam. The roof could then be raised into position by activating the fans, a pressure differential of only 70 mm of water pressure being necessary to maintain it in position. Insulation and weather proofing was to be provided by three layers of roofing felt on 25 mm insulating board, laid with hot asphalt on the sheet steel. All doors were to be double, forming airlocks, except for personnel entrances, consisting of simple

Fig. 9

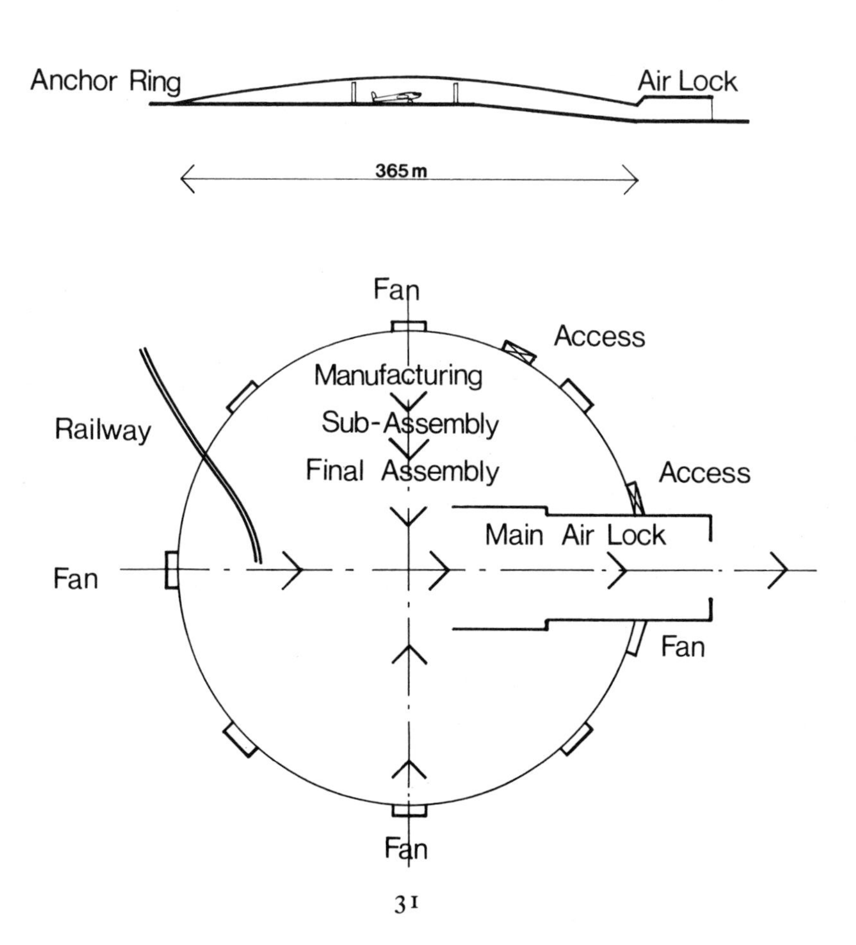

Fig. 9

Plate 8 Barrage type balloons are now used for the transportation of logs

revolving doors. This design, 370 m in diameter, would have completely freed the factory floor of structural obstructions. Stevens claimed that the amount of steel required would be a tenth of the amount used on a more conventional structure of similar size. Like Lanchester's ideas, these proposals were never realised, but they are significant because of Stevens' awareness of the environmental implications of air supported construction.

> 'The use of a reflecting ceiling to prevent heat radiation downwards and collection of exhaust air from the hot strata in the upper part of the dome, will also improve the lighting and assist the cooling system should complete air conditioning be used. The factory is ideally suited for complete air conditioning, and a simple piping system using the pressure in the dome will conduct to the outside, dust and fumes of manufacturing operations or processes that contaminate the air.'[8]

Here Stevens sees the pressurised air supply performing two distinct tasks; not only does it provide stability as a structural element, but it can be treated to provide the required environmental conditions within the building.

Pneumatics for the Second World War

The Second World War was responsible for intensive research on

Plate 9 An air inflatable Centurion tank developed for the British Army for deception purposes

pneumatic structures for certain very specialised applications. Military strategists have always endeavoured to find the ultimate mobility whether it be by land, sea or air. The terms functionalism, miniaturisation and portability are all synonymous with the design of military equipment. These characteristics are, to a large extent, inherent in pneumatic construction, and it is therefore not surprising that military designers, particularly in a time of severe military activity, exploited this form of construction for many applications. Barrage balloons had been used so successfully in the First World War by Britain, that a lot of the countries involved in the Second World War erected them above important military and industrial installations, thus affording protection against low-flying bomber and reconnaissance planes. Air inflated construction was also used to make dummy buildings, weapons and vehicles for deception purposes.

Plate 8

Plate 9

Destruction of air and sea transportation vehicles often resulted in staggering loss of life, and men's main hope for avoiding death lay in the availability of practical survival equipment. The prime requirement of such equipment is that high performance should be instantly available from a small mass. The air inflatable life-raft, occupying little storage space, yet inflated ready for use in a matter of seconds, provided an unmatchable answer. Extensive development has resulted in highly sophisticated air inflatable survival equipment which is now widely used throughout the world.

Plate 10

Plate 10 More lives saved by an inflatable survival life-raft

Radomes; the First Air Supported Buildings

At last in 1946 came the fulfilment of Lanchester's ideas. By then the U.S.A. had developed large radar antennae to protect their northern frontiers against invasion. These delicate antennae had to be sheltered from the severe climatic conditions encountered in these remote areas by some form of enclosure, which had to offer the minimum of obstruction to the radar waves. The Cornell Aeronautical Laboratory, headed by Walter W. Bird, proposed a membrane enclosure supported by air pressure alone, and although the U.S.A. Government of the time was very sceptical about this proposal, the Laboratory was awarded a contract to determine the feasibility of pneumatic structures for this application. In the follow-

Plate 11

Plate 11 The first air supported buildings, one of the many DEW-Line radomes that have withstood well the severe Arctic climate

ing years, after a prototype had been designed, manufactured, erected and tested, over a hundred of these air supported radomes sprung up in the extreme climate of the northern frontier. Of prime importance to the whole project were the Laboratory wind tunnel tests, which analysed the stresses induced in the membrane by wind loads.[9] In association with this research, membrane materials were developed which were able to withstand severe exposure. These consisted of strong man-made fibres, such as nylon or terylene, which were covered with a synthetic coating of vinyl, neoprene or hypalon. As Lanchester had predicted, a pressure differential of only 70 mm of water pressure was all that was required to maintain the rigidity of these 15 m diameter radomes in winds of up to 240 k.p.h. By the mid 1950's, the successful performance of these radomes in the extreme climate of the northern frontier of America had proved the practicality of pneumatic structures. Few ideas have been introduced to the building industry and undergone such severe trials so creditably.

The First Commercial Applications

Due to the success of these radomes, Bird and some of his colleagues decided to form a company concerned solely with the design and manufacture of pneumatic structures, and by January 1956 Birdair Structures Incorporated were in business. Despite Bird's acquired knowledge on pneumatics, CIDair Structures managed to erect the first commercial air supported warehouse, just ahead of Birdair's

Plate 12 Walter Bird's own swimming pool enclosure, with transparent sides

own prototype. This was the signal for great pneumatic activity in the U.S.A. Fabric, tent and parachute manufacturers all started producing simple air supported buildings, and by 1957, there were about fifty manufacturers utilising the portable characteristics of air supported structures, to form enclosures for indoor sports, exhibitions, storage, factories and military applications. The advantage of these structures was that large areas could be economically spanned, for about £5 per square metre of floor area, with a very portable building that could be erected or dismantled in as little as a day. The pneumatic was so novel that it attracted immense publicity and it was not long before the U.S.A. was sprouting hundreds of these pneumatic 'bubbles'. Unfortunately, some of the manufacturers entered the pneumatic field with insufficient knowledge and caution, and consequently several disasters occurred, although luckily without anyone being injured.

Plate 12

Plate 13

This, along with the two major design problems of air supported construction, that of providing a continuous air supply to maintain stability and that of preventing excessive air leakages through entrances, made designers look more closely at air inflated construction for building structures. Air inflated rib construction presented neither of these two problems, but was found to be most suitable for small span structures. For longer spans very high pressures or large diameter tubes were needed. High pressure structures, although their stability was not noticeably affected by pressure

Plate 13 One of Birdair's early structures, the U.S. Army Pentadome that has been extensively used for missile assembly, antenna construction and military exhibits

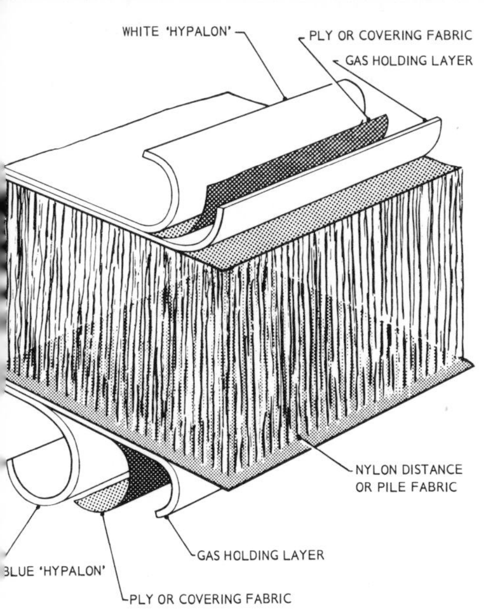

Plate 14 Dual walled airmat type fabric, as used for the M.L. inflatable shelters

variations caused by temperature changes, demanded much stronger and more airtight membrane materials.

Development of 'Airmat' Construction

These shortcomings led to air inflated dual walled construction. Shortly after the Second World War this type of construction was

Plate 15 The M.L. flying air mattress, that could be dismantled and carried in a car luggage compartment

closely investigated both in America and Britain, and in both countries double walled construction held together by closely spaced drop threads was developed. This form of construction, manufactured in a single weaving process and then sealed with an airtight coating, is generally known as 'airmat' construction. The ability to produce complex structural contours with this construction was demonstrated by the Research and Development Establishment at Cardington, England, which in 1951 developed an inflatable aircraft. A similar aircraft was built in America using Goodyear 'Airmat'. Despite its versatility for shaping, 'airmat' construction has distinct limitations. Its complex construction make it very expensive and difficult to repair when damaged, and the internal pressurisation is still quite high, between 20,000 and 80,000 N/m^2, depending on the distance between the walls.

Plate 14

Plate 15

Because of these limitations, Walter Bird carried out a considerable amount of research in developing low pressure dual walled structures, under internal pressures of less than 10,000 N/m^2, and found that quite thick wall sections were needed for stability. Bird used this form of construction in his proposals for a circular exhibition pavilion for the Ford Motor Company. This consisted of inflated circular section panels, running radially on the roof and vertically on the walls, thus following the direction of the major membrane stresses.

Evolution of 'Hybrid' Structures

This intensive investigation into air inflated constructions, however, showed that for larger structures air supported construction was considerably more economical and possessed greater structural

Plate 16 Barracuda air supported shelter enclosing a large ice rink with accommodation for 3500 spectators in Sweden

stability. Consequently designers evolved 'hybrid' structures, in which auxiliary structural elements were used in connection with air supported construction, to safeguard against collapse due to air supply failure, and also to simplify the problem of access. The elements, generally of tubular metal or even inflated ribs, supported the membrane under conditions of low external loading. When these loading conditions increased in magnitude, the interior could be pressurised against this extra loading.

However, economic considerations led manufacturers to concentrate mainly on simple air supported structures, since these undoubtedly held great commercial promise. These structures were usually designed by engineers and were of either cylindrical or spherical form, purely enclosures showing little architectural imagination, although admittedly in their simplest form, the spherical radome, they attained great beauty.

Plate 16

Boston Arts Centre Theatre

It was not until 1959 that architects really began to show an interest in this new structural development. The first architect to use pneumatic construction in a work of architecture was Carl Koch. Aided

Fig. 10

Plate 17 Boston Arts Centre Theatre

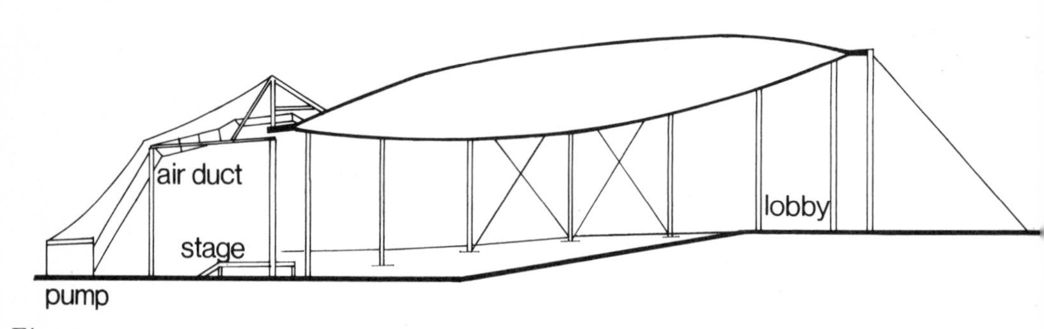

Fig. 10

Plate 17

by engineer Paul Weidlinger, he designed a pneumatic roof for the Boston Arts Centre Theatre. Lack of sufficient money for a permanent structure suggested some kind of tent structure, but no tent design could be found that provided enough free unobstructed area. An air filled, disc shaped roof, 44 m in diameter and 6 m deep at its centre was chosen. Since Birdair had designed and fabricated air inflated radar antennae for military applications, they were the obvious choice for fabricating the roof. The two skins were zipped together and attached to a steel compression ring supported on columns. Two compressors were needed to keep the roof inflated, and the temperature of the air supply was controlled to suit climatic conditions. The whole roof was tilted, to take advantage of the acoustic properties of the lower convex surface. This is the first example of a controlled pneumatic environment.

Plate 18 Portable Exhibition pavilion for the United States Atomic Energy Commission

Lundy's U.S.A.E.C. Portable Exhibition Pavilion

Closely following this, in 1960, the architect Victor Lundy achieved another major breakthrough in pneumatic structuring. His portable

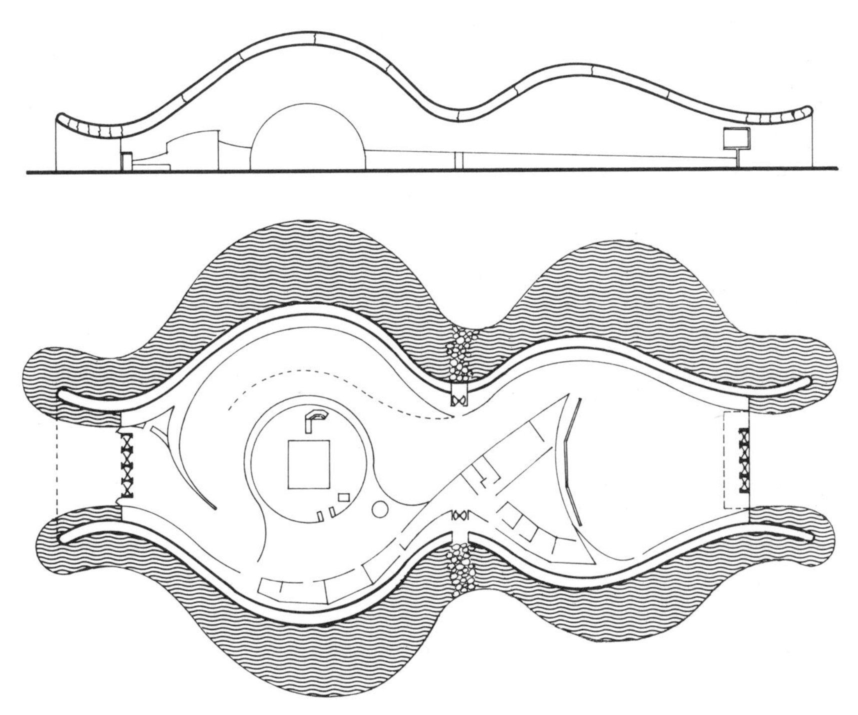

Fig. 11

Plate 19 U.S.A.E.C. portable exhibition pavilion, entrance

Fig. 11 exhibition hall for the United States Atomic Energy Commission, again manufactured by Birdair with Severud as consulting engineer, remained the only truly sophisticated pneumatic architectural state- *Plate 18* ment throughout the sixties. The A.E.C. specifications were for portability, low cost, and safety, and Lundy's solution was an air supported structure 90 m long with a maximum width of 38 m and a maximum height of 15 m, consisting of two domical forms connected with a saddle junction. It was only after considerable research on various building types that Lundy came to the conclusion that an air supported structure presented the best solution. The building was constructed of two vinyl-coated nylon membranes with a 1·2 m air space between, divided into eight compartments so that accidental damage to any one part would not affect the overall stability of the building. The inner and outer membranes were tensioned by a pressure differential of 49 mm and 38 mm of water pressure respectively. This internal pressurisation was chosen to withstand wind loadings of up to 150 k.p.h., but it safely withstood a wind loading of over 200 k.p.h. at Santiago, Chile. The air space between the membranes provided adequate insulation, avoiding the need for cooling equipment. Entry was by means of revolving doors in a rigid frame at either end of the building. For bulky equipment, the two *Plate 19* air inflated entrance canopies at these ends, served as an airlock, whilst such equipment was moved in and out of the building. The structure could be erected in three to four days by a team of twelve *Plate 20* labourers under competent supervision. No special equipment was

Plate 20 U.S.A.E.C. portable exhibition pavilion, the erection procedure

Plate 21 U.S.A.E.C. portable exhibition pavilion, interior view

required, and within 30 minutes of the pressurisation fans being activated the structure was fully inflated. Lundy's research suggested that this erection time was one-fifth to one-twentieth of that of other building structures that were considered. The total weight of the building, including all equipment, such as pressurisation fans, was less than 30,000 kg, and when packed ready for moving on to the next site, its volume reduced to less than 150 m^3. The building's cost was slightly less than £20 per square metre of floor area although at each new site quite considerable site preparation was often necessary, sometimes costing as much as £10 per square metre. Despite its external rather obese dominance, internally the building played a subordinate role to the exhibits, due to the use of dark blue lighting around its periphery, whilst the exhibits were highly illuminated. The building structure housed a theatre with seating capacity for 300 people, a technical laboratory and also lecture areas. Amongst the exhibit was an experimental atomic reactor housed in a small transparent air supported dome—a bubble within a bubble. Certainly as an exhibition enclosure this building was most successful, but what is more significant is that here architecture for once emerged out of the exploitation of a new technology. This building, along with the Boston Arts Centre Theatre, clearly demonstrated to the architectural profession that pneumatics were far more than mere temporary structural enclosures.

Plate 21

European Air Supported Beginnings

It was not until the late 1950's and the early sixties that European designers took a serious interest in the pneumatic building developments occurring across the Atlantic. The public first became aware of it through the Pan American Airways Pavilion at the 1958 Brussels World Fair. This was a large globe structure, which showed the extent of the company's air routes, and was surrounded by a reinforced concrete viewing gallery. Pneumatic activity followed in Europe, particularly in Germany and Sweden, and although the United States of America led the way technologically, it was the Europeans that took over the lead given by Koch and Lundy. The greatest of these is undoubtedly Frei Otto, the foremost figure in the field of the tensile structures. In his search for structures that required a minimum of time and material, he became intensely interested in pneumatic construction. Following considerable research, he produced a breathtaking publication which probed into general principles and presented an abundance of imaginative ideas on pneumatic forms and applications.[10]

Development of the Ground Effects Machine

Although pneumatic building construction was rare in Europe during the 1950's, another type of pneumatic structure was being developed in Britain and this could perhaps have a great impact on future transport systems and even on future living patterns. This, the ground effects machine, G.E.M., although more generally known in Britain as the hovercraft, is an air controlled structure, which moves on a cushion of air contained between the vehicle and the surface over which it is operating. This cushion of low pressure air must be sufficient to support the vehicle's weight. Basically three distinct types can be distinguished. Firstly, the air bearing hovercraft consists of a flat plate with a hole in its centre through which air is forced between the plate and the operating surface. Secondly, the plenum chamber hovercraft has a concave surface under structure in which the air is pressurised sufficiently for uplift purposes: the air leaks out under a flexible skirt. Thirdly, the momentum curtain hovercraft derives its support from air jets around the vehicle's periphery, and

Fig. 12

Fig. 13

Fig. 14

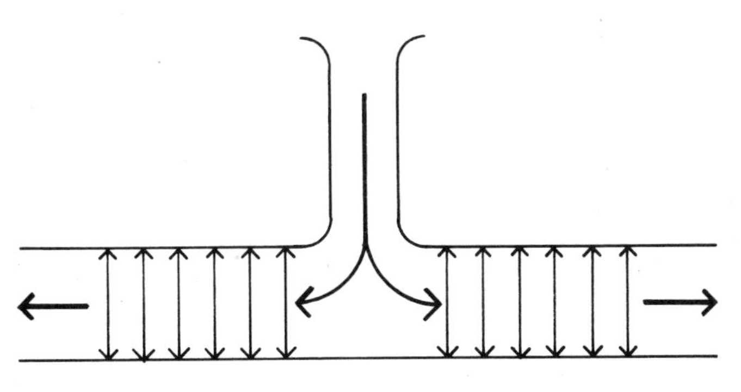

Fig. 12

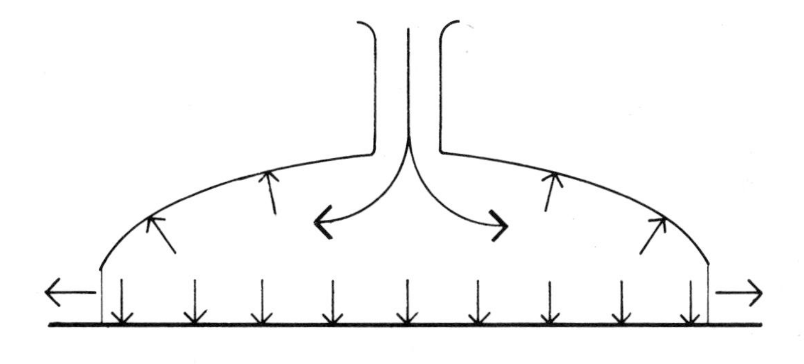

Fig. 13

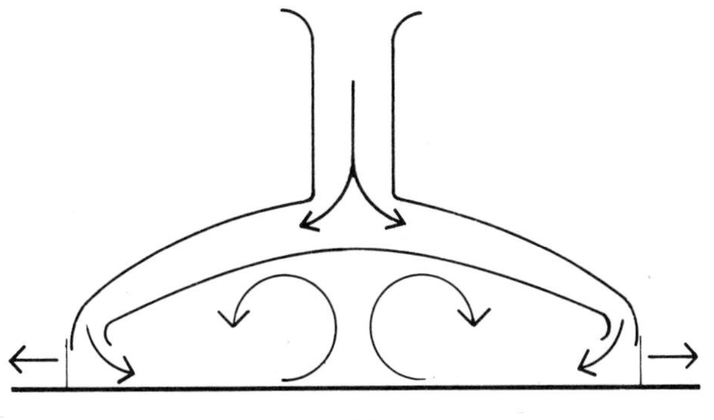

Fig. 14

Plate 22 In 1969 the first regular hovercraft service was in operation across the English Channel

these are directed both inwards and downwards. With additional recirculation jets in the centre, the efficiency of the momentum curtain hovercraft can be considerably improved. Although the pioneering development of the hovercraft was conducted in Britain, many other countries are now engaged on hovercraft development. Already there is a regular hovercraft service across the English Channel, which despite teething troubles has been very successful. Here again the novelty of the pneumatic structure has attracted much public attention. This type of vehicle frees transport from the fixed routes dictated by road and rail, and perhaps in the future could be the vehicular basis for a completely go-anywhere, mobile living unit. It is this possibility that makes a mention of the ground effects machine very relevant in this context.

Plate 22

Pneumatics in Art

During the sixties pneumatics also became popular with the 'Kineticists' as an art medium. Leonardo da Vinci was perhaps the earliest artist to have realised the aesthetic beauty possible with air, with his creation of a pneumatic environment using inflated pigs' bladders.[11] Marcel Duchamp and Picasso both experimented with this medium at the beginning of this century. The 'Structures Gonflables' Exhibition at the Musée d'Art Moderne in Paris, March 1967, and the Exhibition of 'Air Art' at the Contemporary Art Centre, Cincinnati, Ohio, U.S.A., March 1968, heralded the undoubted acceptance of air as an artistic medium. Such works of art give

Plate 23 Air inflated roof over the 'Brass Rail' restaurant pavilions at the New York World Fair 1963–64

designers the much needed enlightenment and introduction to the aesthetic characteristics of pneumatics for non-fine art experiences. In addition to this, the Baschet Brothers have performed with pneumatic, musical sculptures, which in their own words are a 'synthesis of shape, sound and human participation', and these undoubtedly give designers some insight into the acoustic characteristics of a pneumatic environment.

World Expositions Increase Pneumatic Enthusiasm

New ideas on building techniques often figure prominently at world expositions, and indeed pneumatics are no exception. Following the unveiling of the pneumatic Pan American pavilion at the 1958 Brussels World Fair, several pneumatic pavilions were constructed for the New York World Fair 1963–1964. Surprisingly these pavilions evoked little interest and reaction despite the fact that the inflated roofs to the 'Brass Rail' restaurant pavilions are amongst the most dramatic pneumatic structures yet to be conceived. Like the A.E.C. exhibition pavilion, these were again the result of collaboration between Victor Lundy, Severud and Birdair. These structures conceived for a carnival atmosphere, demonstrate the sculptural excitement possible with pneumatics. Each inflated roof, 23 m high and 18 m in diameter, was a sculptural container, anchored to a central mast so that it hovered over the refreshment area. At night time, lighting within this volume produced a glowing form, in which the intricate fabrication patterning was revealed. Here architecture was alive, for a full 24 hours of every day. In complete contrast to the New York World Fair, EXPO '70 has been the greatest pneumatic happening ever, and its pneumatic pavilions, which range from the sublime to the ridiculous, have given pneumatic structuring a much needed impetus. Now enthusiasm for pneumatic buildings, for pneumatic furniture, and even for 'pneumatic happenings' is rapidly expanding, not only amongst designers but also the general public.

Plate 23

Plate 24

The First International Pneumatic Colloquium

Despite the great prominence of pneumatics at EXPO '70, its significance in pneumatic history is certainly no greater than an event that occurred 3 years earlier on the 11th and 12th May 1967; this, the International Association for Shell Structures Pneumatic Colloquium, was held at the University of Stuttgart in Germany.[12] Perhaps the greatest achievement of this Colloquium was to stimulate the participants' awareness of the huge gulfs separating architectural designers, engineers, manufacturers and mathematicians.

> 'I discovered that most of the part of our family dealing with reality cannot understand the other part dealing with theory. Many of you know that I am striving very hard to introduce more basic theory in this field and to stimulate mathematicians to discover it. I feel it is my task to be an interpreter between both these groups, however, I too get lost. I wish that the theoreticians fly as high as possible; but they must describe such flights with words that the earthbound practitioners in our family can understand. On high flights both the control tower on the ground and the pilot himself must know the heights. We must attempt to overcome this gap of misunderstanding, but without compromising on essentials.'[13]

These concluding words of Frei Otto have fostered some communication between pneumatic designers, which is undoubtedly leading to a clearer understanding of pneumatics. However, this exchange of ideas between designers, manufacturers, users and legisla-

Plate 24 A 'Brass Rail' restaurant pavilion, an illuminated jewel at night

tive bodies could still be vastly improved. Research and development in both design and manufacture of pneumatics still tends to be isolated, particularly in the latter, where rivalry between commercial competitors encourages a certain amount of secrecy concerning constructional techniques and manufacturing methods. This guardedness and lack of communications is most certainly detrimental to the progressive development of pneumatic building construction. It is now up to designers to overcome these problems so that this new building form can be exploited to perhaps achieve an architecture more compatible with present-day technological development.

REFERENCES

1. B. Clarke, *The History of Airships*.
2. C. Dollfus, *Balloons*, p. 36.
3. B. Clarke, op. cit., p. 22.
4. F. W. Lanchester, Patent 119,339, 1917.

5. F. W. Lanchester, *Span*, Manchester Association of Engineers, Lecture, p. 28.
6. F. W. Lanchester, op. cit., p. 34.
7. H. H. Stevens, 'Air Supported Roofs for Factories', *Architectural Record*, December 1942.
8. H. H. Stevens, 'Air Supported Roofs for Factories', quoted by permission from the December 1942 issue of *Architectural Record*, copyright 1942 by McGraw-Hill Inc., with all rights reserved.
9. For further details see the following Cornell Aeronautical Laboratory Reports: UB-909-D-1; UB-909-D-2; UB-512-J-1; UB-664-D-1: and UB-747-D-19.
10. F. Otto, *Zugbeanspruchte Konstruktionen*, Band 1.
11. W. Sharp, 'Air Art', *Architectural Design*, March 1968, p. 99.
12. *Proceedings of the 1st International Colloquium on Pneumatic Structures*, University of Stuttgart, 1967. Copies obtainable through Institut für Modellstatik, Technische Hochschule, 7 Stuttgart, Postfach 560, Germany.
13. F. Otto, *Proceedings of the 1st International Colloquium on Pneumatic Structures*, p. 179.

3. An Analysis of Pneumatic Behaviour

Since pneumatic construction is still in its early days as an architectural technology, it is not surprising that a comprehensive understanding of its complex behaviour has yet to be achieved. It possesses an individuality that is far removed from conventional structural techniques, and is clearly apparent when some of the parameters which influence pneumatic forms are considered. For instance the membrane tensions of an air supported dome structure under the surface loading of internal pressure only is indicated by the simple expression

$$T = \frac{P_i R}{2} \qquad (3.1)$$

Fig. 15

If two domes are considered, which cover equal areas and are equally pressurised, but the radius of curvature in one is twice that of the other, then obviously, the membrane stresses in the shallower one will be twice as great as in the other. Thus the membrane material and the anchorages must be of the order of twice the strength. However, the shallower dome is more streamlined, forming less of an obstruction to the wind. This means that smaller stabilising pressure

Fig. 16

Fig. 17

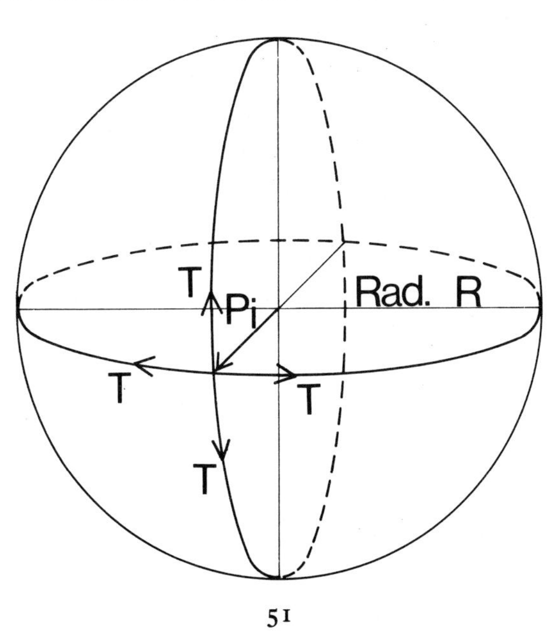

Fig. 15

differentials are required in the shallow dome. When external surface loadings, such as wind and snow, are considered, this apparently simple mathematical expression becomes much more involved. Consequently, the determination of the most suitable structural engineering form needs considerable study. In addition to this, the environmental parameters are of considerable influence; vast volumes of air must be conditioned, the thin membranes tend to cause condensation, and the curved surfaces, characteristic of pneumatics, need careful acoustical consideration. All those parameters, not to mention aesthetics, can each have a very distinct influence on the form of a pneumatic, and this suggests that sophisticated pneumatic design can only be the result of an integrated dialogue between architects and engineers.

The following pages look at the influencing parameters that dictate pneumatic behaviour, what these parameters are, and also what their influence is. This discussion is not intended as a detailed manual for pneumatic design, but as an introduction which gives a basic understanding of pneumatic behaviour and acts as a stimulant for further theoretical work.

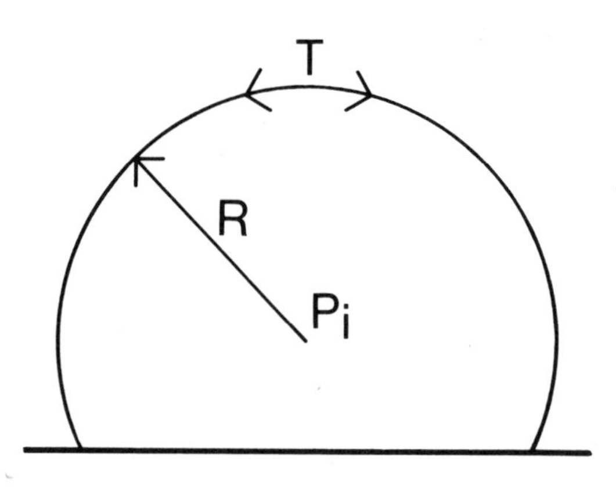

Fig. 16

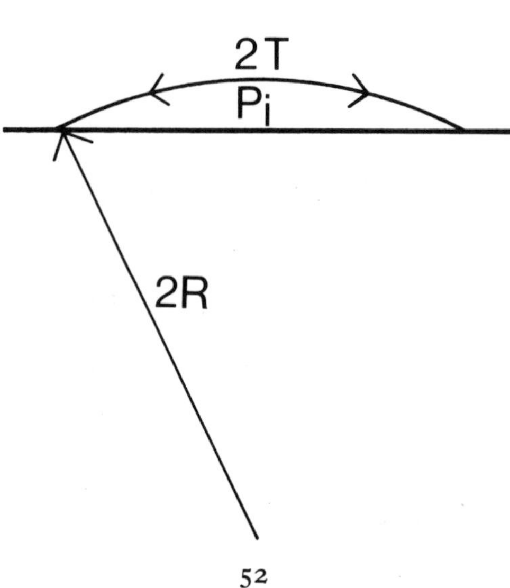

Fig. 17

To clarify the understanding of complex phenomena, analogies are often used. With pneumatics we need not search very far for an appropriate analogy, for soap bubbles and films are one of the purest forms of pneumatic structure. Much can be derived from studying their behaviour[1]; indeed Frei Otto, in his search for structures utilising a minimum of material and time has conducted many experiments with them.[2]

A soap bubble is moulded by the surface tension forces acting on both sides of the soap film. Due to the uniformity of these forces, the main characteristic of the film is to form shapes of minimal surface area, in which the walls are stressed equally at every point, and in all directions, with no concentration of stress at any one point. Stresses are equalised by liquid flow in the soap film, so that stress peaks can in no circumstances occur. Any shape that can be achieved with a soap film, is suitable for pneumatic construction. If a membrane is made out of extremely inelastic material and exactly shaped in the form of an enlarged soap bubble, this membrane, provided its weight is negligible, will also be uniformly stressed at every point and in all

Plate 25

Plate 25 A freely floating soap bubble

Plate 26 A hemispherical captive soap bubble

directions when inflated. Soap bubbles are acted upon by internal pressure loads only, and consequently, deviation from soap bubble forms may be necessary to obtain the most suitable shape for pneumatic structures, which must also be capable of resisting external loading. The smallness and fragility of soap bubbles make it difficult to experimentally analyse their form and subject them to external loading conditions. Despite this, experimentation with soap films is most valuable in the early design stages, since it can suggest to the designer, suitable pneumatic forms which may accommodate the design parameters within which he must work.

The variety of pneumatic forms possible is clearly demonstrated by inflating soap films over various complex plan shapes. With all such inflated soap films, the minimal surface area shapes obtained have a clearly discernible tendency to generate spherical forms, doubly curved synclastic forms being usual, although singly curved and anticlastic forms are by no means impossible. Besides this tendency towards minimal surface areas, soap bubbles conglomerate in a certain manner; soap films always meet at an angle of 120° to one another and the relationship between the radius of curvature of the film and the pressure differential across it is always constant. Thus in the floating twin bubbles of equal size, the dividing wall between them is planar since the pressure in each bubble is equal. With

Plate 26
Plate 27
Fig. 18

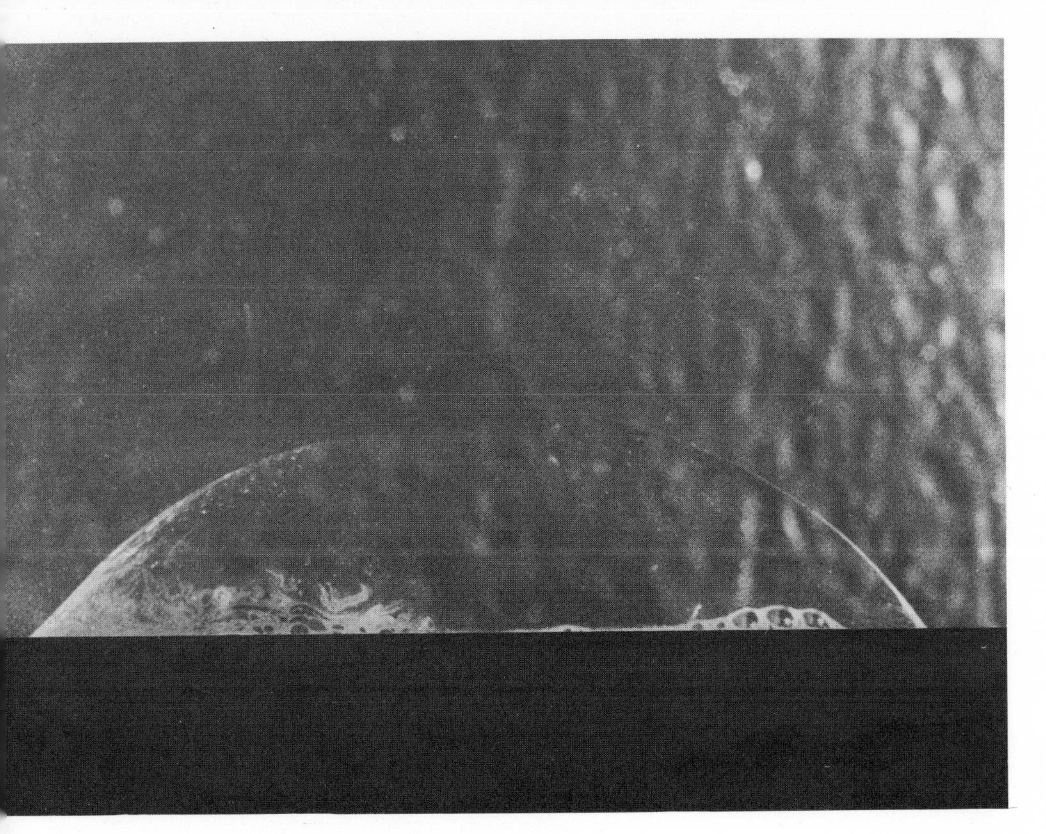

Plate 27   Soap bubble on a rectangular base

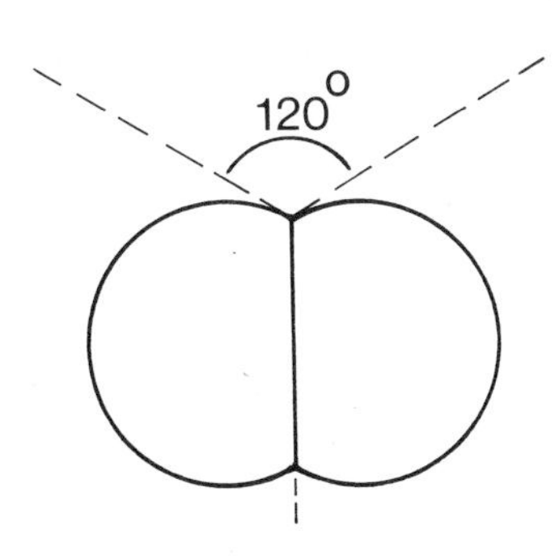

Fig. 18

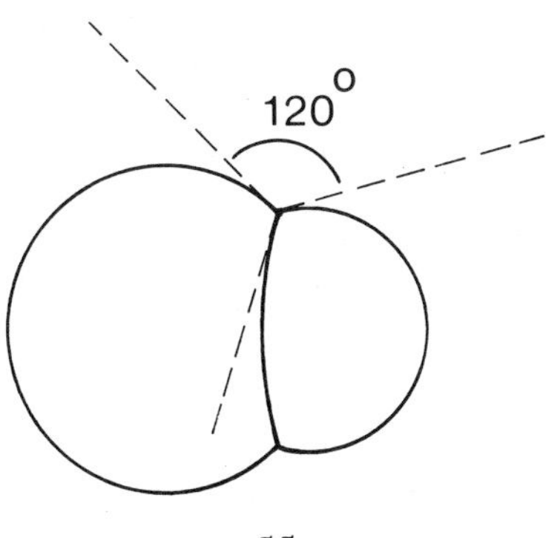

Fig. 19

Plate 28 Soap bubble conglomerations

bubbles of different sizes, their internal pressures are different, and mould the common membrane wall to a curved form; however, the tangential angles between the soap films remain 120°. No matter how many bubbles are grouped together, the soap films always meet at the same angle. Although a fixed number of bubbles can group themselves in a multitude of arrangements, there is an observed tendency for these bubbles to adopt circular plan forms, those in which a maximum plan area is enclosed with a minimal membrane surface area. These bubble conglomerations could perhaps be taken as architectural prototypes for pneumatic buildings. The large bubble surrounded by numerous smaller bubbles suggests a pneumatic form of perhaps an assembly hall with ancillary accommodation such as foyers, restaurants, bars, conference rooms, toilets and cloakrooms or even a huge aircraft hangar with its many adjacent offices, stores and workshops. These arrangements also suggest ways in which pneumatic buildings can be partitioned at will by internal membrane walls and indicate how these inner walls can influence the structural form of pneumatics. It must be emphasised again, that such studies can only serve as a guide to the understanding of pneumatics and make it possible for the designer to examine the various pneumatic forms that might fulfil his functional and aesthetic design parameters.

Fig. 19

Plate 28

FREI OTTO'S PNEUMATIC LAWS OF FORMATION

In addition to this, Frei Otto has also examined the laws that govern pneumatic forms.[3] He came to the conclusion that any shape generated by revolving a linear form about an axis could be pneumatically achieved. Consequently by use of the inscribed sphere it can be ascertained whether a certain form is suitable for pneumatic construction. When this law of formation is applied to a cylindrical membrane it defines the form as a series of equal sized spheres, an infinitely small distance apart. These laws of formation, although setting considerable limiting factors, extend the possible variation in form far beyond the simple spherical and cylindrical shapes generally employed for pneumatic structures.

Fig. 20

AIR SUPPORTED CONSTRUCTION

Design Loadings

As the behaviour and characteristics of this form of construction are so unique compared with those of air inflated construction, more attention in this study has been given to the analysis of air supported structures. As they possess such dynamic characteristics, the influence of loadings is more clearly perceived than with any static, more conventional structure. The loading conditions are very similar to those encountered by any other structure in comparative circumstances, but with one major and essential addition, the pressurisation load. It is this one load that reacts against and supports all the other loads. Consequently, in air supported construction there are three loading types to be considered, dead loads such as structural self-weight and loads suspended from the membrane, live loads caused by climatic parameters such as snow, rain and wind, and finally the pressurisation loads.

Self Weight Loadings

The self weight of the structural materials, at present employed in pneumatic construction, is so insignificant, compared to other applied loads, that it can usually be neglected for design purposes. However, such is technology, that the synthetic-coated fabrics generally asso-

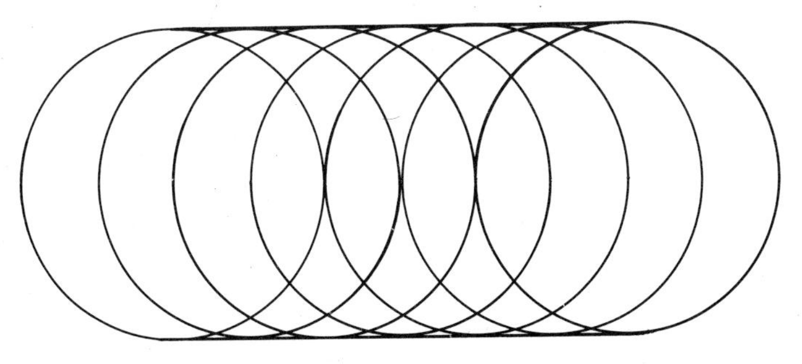

Fig. 20

ciated with pneumatics will not always be employed. Heavier materials with perhaps greater strength, perhaps greater permanence or even better insulating characteristics may well be used in the future, and when these are employed, their self weight will indeed be of greater significance.

Considering a semi-cylindrical shaped structure, with a membrane material whose self weight load intensity is 'G', the total loading across the membrane, at a point, is given by:

$$P = P_i - P_g$$

Fig. 21 where

$$P_g = G \cdot \cos\theta$$

$$\therefore \; P = P_i - G \cdot \cos\theta$$

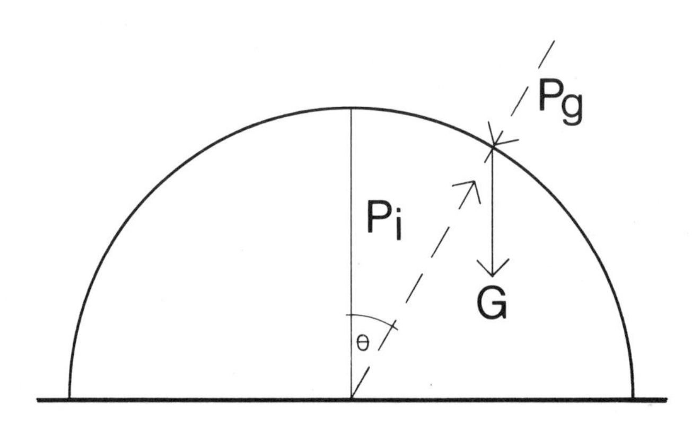

Fig. 21

From this expression it can be seen that the influence of the self weight is greatest on the horizontal plane and decreases gradually to zero as the vertical plane is approached. Consequently the structure will be deformed from its cylindrical shape, taking on a flatter curvature at its crown. Even with the lightweight materials generally used for the air supported membrane, a very slight flattening effect can still be detected. *Fig. 22*

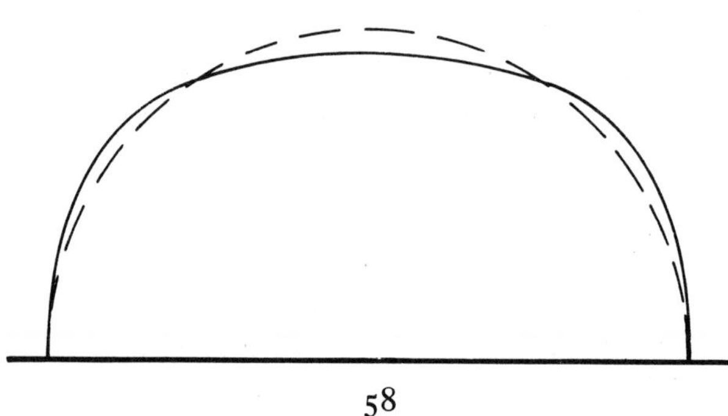

Fig. 22

Point Loadings

In addition to the self weight, other types of loading may be applied to the membrane. These are often in the form of concentrated point loads which not only deform the membrane quite substantially, but also cause severe local stressing of the membrane. Although experience has shown that air supported membranes can carry some concentrated loading (it is possible for a person to walk over an air supported structure without deforming it very noticeably), little is known about the relationships between load magnitudes, internal pressures, the extent of deformation, induced stresses and membrane material characteristics. Some research has been carried out in Russia, but this has only considered the influence of the internal pressure on deformation.[4] Where this type of loading is unavoidable, it is advisable to adopt the following procedures; either the resulting induced stresses must be distributed over as large an area as possible of the membrane, or the loads must be carried by reinforcement elements, such as cables, nets or even a thickening of the membrane, but in both cases this applied loading must not cause the development of compressive stresses in the membrane.

Fig. 23

Snow Loadings

Like concentrated point loads, snow presents a loading problem about which very little is known. Although simulation tests on snow loading can be carried out, the true picture of pneumatic behaviour under snow loading can only be obtained by observations of reality. In no two cases is the magnitude and influence of snow loading the same. The magnitude is dependent, firstly on climatic conditions, and secondly on the individual characteristics of the structure, its curvature and form, its stability, its environmental standards and its insulation properties. With the simple spherical and cylindrical forms at present utilised in pneumatic construction, the snow load is not usually a major problem and in some cases it can be ignored altogether. Due to the smoothness of the membrane material, the snow tends to slip down the sides of the structure and only accumulates at the crown. In many cases, even this accumulation can be

Plate 29

Fig. 24

Fig. 25

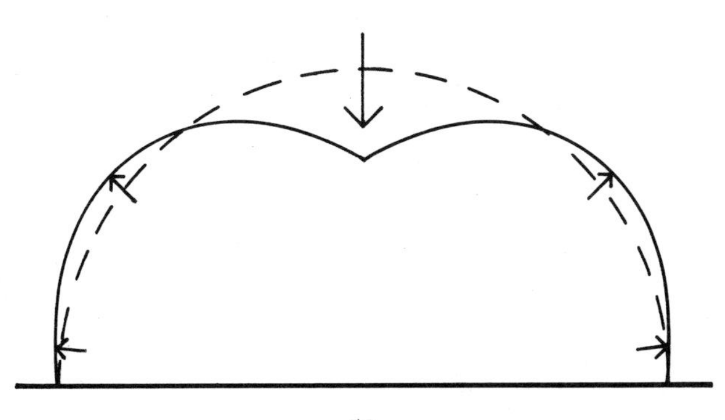

Fig. 23

Plate 29 The snow does not remain long on these three air supported warehouses fabricated by Barracuda

avoided by a change of pressurisation load, which effects a relaxation and then a retensioning of the membrane. This relaxation causes the snow's adhesion with the material to be broken, and when the structure is re-inflated, the expanding movement forces the snow to slide off. A cruder but nonetheless effective way of breaking this adhesion, is to beat the sides of the structure, or to drag a rope over it. Inflatable tubes distributed over the membrane's surface or even electrically heated wires embedded in the membrane material are more sophisticated methods for preventing snow adhesion, but these are yet to be employed. If a comfortable internal environment is maintained, with

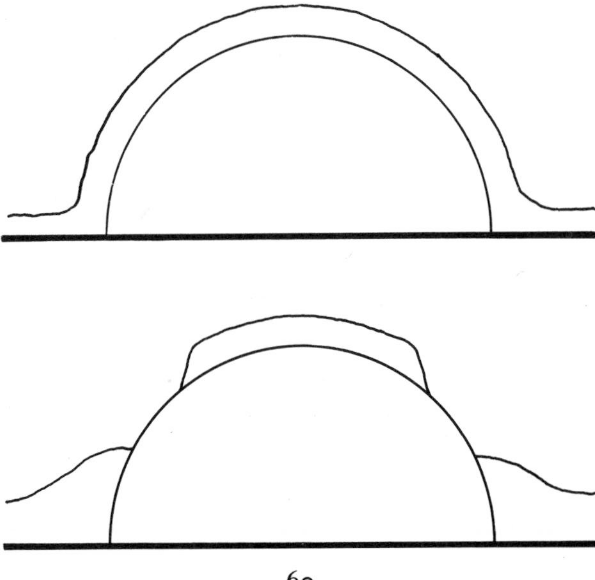

Fig. 24

Fig. 25

regard to temperature, and membrane materials of negligible insulation are employed, thawing of the snow will occur to some extent, reducing the adhesion and so causing the snow to slide off. In some countries the building authorities allow designers to ignore the snow load, provided that a specific minimum internal temperature is guaranteed within the structure,[5] but such regulations are perhaps rather dangerous. As soon as the snow loads start to cause deformation in the curvature of the structure, severe problems arise. The flattening of the structure attracts further snow accumulation which causes greater deformation, and this will possibly lead to eventual collapse of the structure. When the influence of the wind on these snow loads is considered, the problems become even more complex, since drift formations must be taken into account.

Fig. 26

Snow-drifting is itself a subject about which research has only recently begun. Drifts are caused not only by fresh snow falls, but also by the movement of recently deposited snow, which the wind plucks from the ground, and redeposits in different locations. The wind is capable of carrying a specific amount of snow, which is related to its velocity. With an increase in velocity, the wind transports more snow particles, and once it is fully saturated to its maximum carrying capacity, a reduction in velocity will cause deposition of the snow particles which it is no longer capable of carrying. Consequently, natural snow-drifts occur when wind velocities drop, or during periods of snow precipitations. In addition to this surface structures also cause drifting, because of, firstly, the adhesion of the snow to the structures, and secondly the wind velocity reductions caused by the structures. Any surface obstruction will force the wind over and around it, constricting the volume of air carrying the snow

Fig. 27

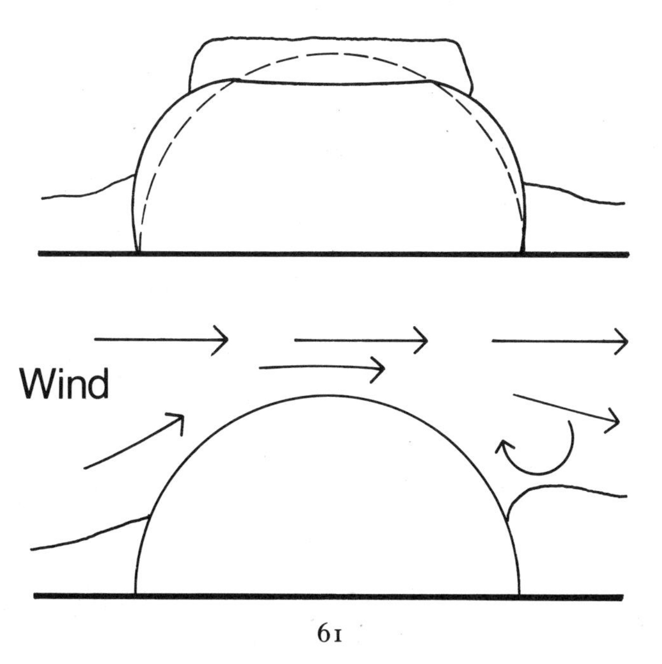

Fig. 26

Fig. 27

and increasing its velocity, and hence its carrying capacity. Having passed the obstruction, the air assumes its original volume, no longer having its added carrying capacity, so that deposition occurs, causing snow build-up on the leeward side of the structure. With snow falls during stormy winds, it is likely that snow will accumulate around the structure and not upon it.

From the above discussion, it is obvious that forecasting the magnitude of the snow loads for design analysis is extremely difficult. However, one approach, although simple and certainly not precise, is perhaps valid. The eventual build-up of snow around the structure is certainly not as significant with regard to deformation, as the initial snow accumulations upon the structure, and it is this latter loading condition which should bear closer examination. Obviously with a curved form, there is not a uniform snow load intensity 'S' over all parts of the structure. The snow loading at a point is dependent on the gradient of the membrane material at that point, with limiting values of 'S' on a horizontal plane and zero on the vertical plane.

Considering a semi-cylindrical shaped structure under a snow load intensity of maximum value 'S', and assuming negligible membrane weight, the total pressure loading intensity across the membrane at a point is given by:

$$P = P_i - P_s$$

where $P_s = S_\theta . \cos\theta$

Fig. 28 and $S_\theta = S . f(\theta)$ between limits $0 \leqslant \theta \leqslant 90°$.

This can be approximately represented by:

$$S_\theta = S . \cos\theta$$

$$\therefore \quad P = P_i - S . \cos^2\theta \qquad (3.3)$$

The ratio of the snow loading intensity to the internal pressure loading has not yet been thoroughly investigated, but some research

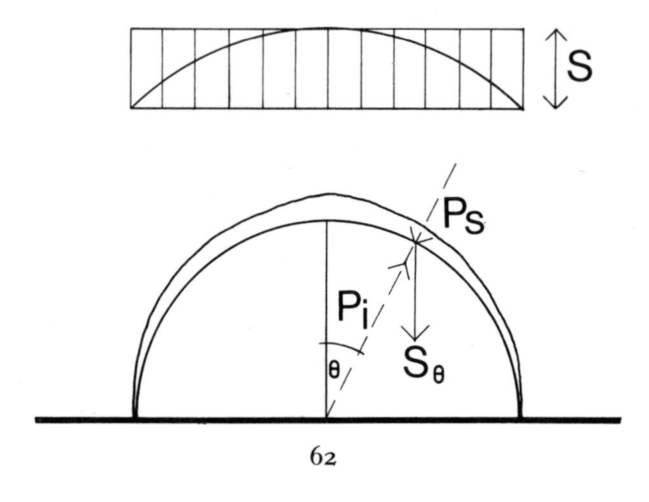

Fig. 28

work undertaken in East Germany suggests that deformations caused by snow loading can be more pronounced than those due to wind loading.[6] Where pneumatic forms do not possess their usual tendency to prevent heavy snow accumulation, the following condition to avoid folding of the membrane will undoubtedly apply:

$$\frac{P_i}{S} > 1 \tag{3.4}$$

However, for the more popular cylindrical and spherical forms, a folding condition less than unity will be appropriate, its exact magnitude being dependent on form, environmental conditions within, and membrane insulation characteristics.

Wind Loadings

When a dynamic structure, such as a pneumatic, is under the action of such a variable force as the wind, it is imperative that the nature of the wind loading is understood. Some designers have utilised the much accumulated knowledge of wind behaviour on rigid structures, and accepted them as valid for pneumatics. However, wind tunnel tests have shown that these design loads, applicable to rigid structures, are only approximate for pneumatic structures, and indeed inaccurate when large deformations of the pneumatic occur. Although the wind causes positive pressure over some zones, the major part of the membrane's surface is subjected to negative or suction pressures. In shallow profiled structures, only wind suction forces act

Fig. 29

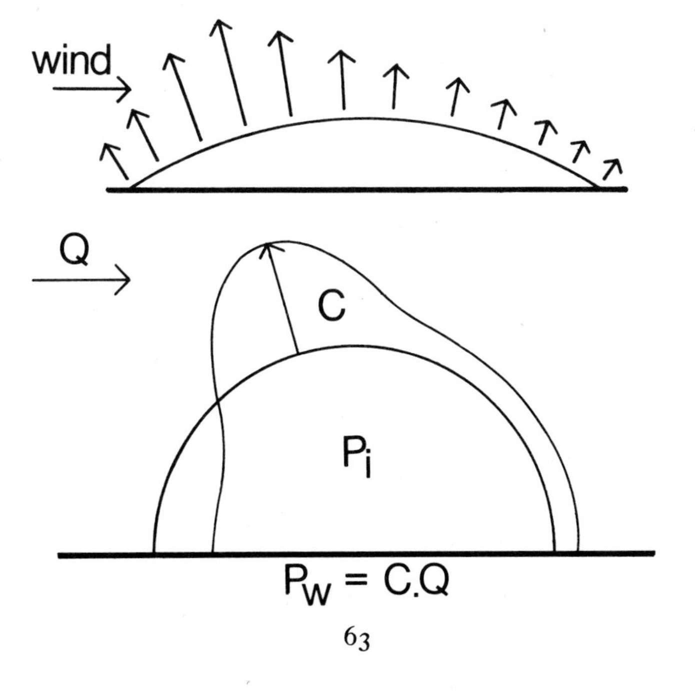

Fig. 29

Fig. 30

upon the membrane, and these may be sufficient for support. The lower the profile of a structure, then the greater stability it possesses, and the less the internal pressurisation load required for supporting purposes.

The wind loading intensity at a point 'P_w' is assumed to act perpendicular to the membrane surface under consideration and is given by:

Fig. 30

$$P_w = C.Q \tag{3.5}$$

where 'C' is the aerodynamic coefficient at the point and 'Q' is the dynamic pressure of the wind acting horizontally. This dynamic pressure 'Q' is dependent on the wind velocity 'v', and the air density 'ρ', as given by:

Fig. 31

$$Q = \tfrac{1}{2}\rho . v^2 \tag{3.6}$$

The aerodynamic coefficient depends on the shape of structure, the wind direction, and the position on the membrane surface of the point under consideration. Quite a few wind-tunnel tests have been undertaken to establish values for this coefficient and there have, in some cases, been quite significant discrepancies in the results. These discrepancies are not surprising when the problems of pneumatic wind-tunnel testing are considered. In practice, membrane materials are generally so light in weight, that when reductions in scale are made for the purpose of model analysis, the resulting materials are extremely flimsy and delicate. This means that any measuring devices will have a pronounced influence on the behaviour of the model, tending to stiffen it and reduce the magnitude of deformations and induced stresses. To prevent this, large models and wind-tunnels are necessary.

A stability criterion must be adopted so that the suitability of

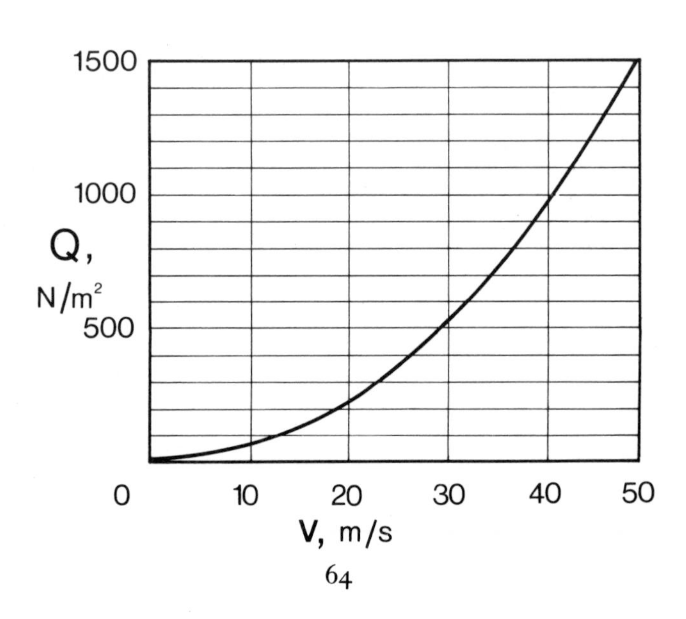

Fig. 31

different forms can be compared. This criterion is that folding conditions on the windward side of the structure are avoided, if the pressurisation load is greater or equal to the wind pressure loading intensity, normal to the membrane at that particular point, as represented by:

$$P_i \geqslant P_w \tag{3.7}$$

Results of various wind-tunnel tests have not been entirely consistent, there being marked discrepancies in the magnitude of the aerodynamic coefficient. The values measured by Bird and the Cornell Aeronautical Laboratory appear to be the most reliable, and until further tests have been made and more accurate and consistent results achieved, it would be most prudent to use these values.[7]

For the three-quarter spherical forms commonly used for radome structures, the folding condition is found to approximate to unity, as given by the condition:

$$\frac{P_i}{Q} \approx 1 \tag{3.8}$$

Fig. 32

Peak values for the aerodynamic coefficient occur in the suction zone and are displaced towards the windward side of the structure.

Fig. 33

With regard to semi-cylindrical and hemispherical forms, wind-tunnel tests in East Germany achieved rather surprising results.[8] The spherical form is regarded as one of the most rigid and efficient structural forms. However, the East German investigations showed that the semi-cylindrical form was more stable than the hemispherical form, the folding conditions being approximately 0·6 and 0·7 respectively; logically the reverse would be expected. It is quite possible, though, that the semi-cylindrical form derived considerably more stiffness from the measuring devices. It would therefore seem reasonable to take the following folding condition for both semi-

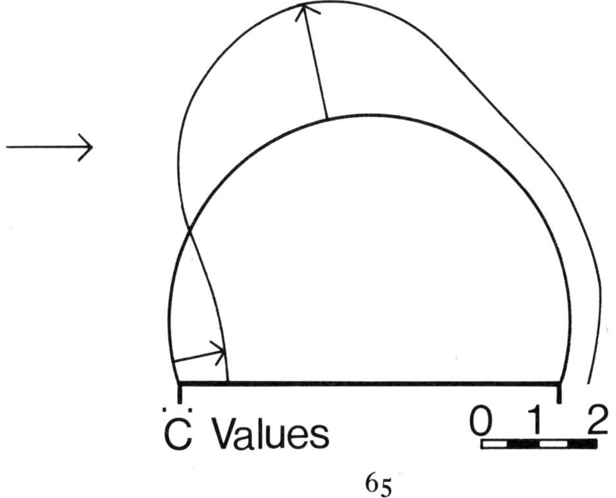

Fig. 32

cylindrical and hemispherical forms, until further tests prove otherwise.

Fig. 34

$$\frac{P}{Q} \approx 0{\cdot}65 \qquad (3{\cdot}9)$$

From existing data derived from wind-tunnel test results, the following conclusions can be drawn for the better understanding of pneumatic behaviour under wind loading.

1 As the profile of the form decreases, so does the condition for folding given by the ratio P_i/Q.
2 Peak loadings occur in the suction zones and are slightly displaced to the windward side.
3 Deformations greatly influence the pressure distribution on the windward side, but are much less significant on the leeward.

Fig. 35

During his wind-tunnel investigations Bird observed that with spheres having base angles 'α' of small magnitude, instability often occurred under high wind loads.[9] This prompted him to undertake further investigations, which revealed that this instability happened when the natural frequency of the structure coincided with the vortex frequency of the wind. As the base angle was decreased, so the critical wind velocity, at which this dynamic instability was present, decreased. This buffeting can be a serious problem with tower mounted radome units, but otherwise does not cause concern with the more popular pneumatic forms.

In this discussion the wind load has been considered as a static load, of constant maximum value for design purposes, but actually,

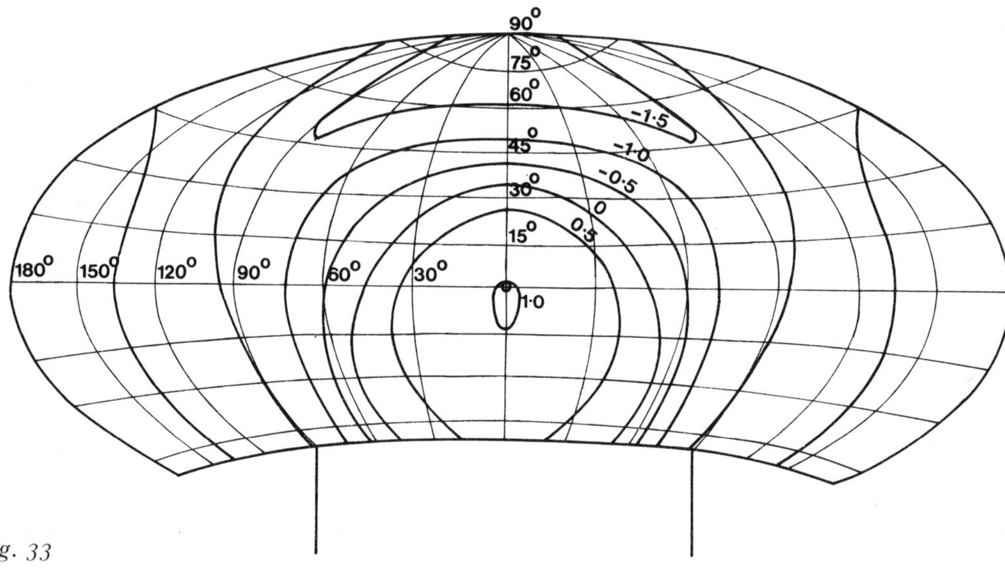

Fig. 33

it is indeed a dynamic load, constantly varying. How then does the pneumatic react to these variations?[10] A sudden increase in the wind load cannot be immediately compensated for by an increase in internal pressure. Therefore the structure will deform, decreasing the volume of air within the structure, and thus increasing the internal pressure as approximately indicated by Boyle's law.* This pressure increase provides added rigidity to the structure against further deformation. However, if these gusts are prolonged, and the internal pressure within the structure is automatically regulated to maintain a constant level, the internal pressure soon returns to its normal level. When eventually the wind gust diminishes, the structure reverts to its original form of increased volume and this causes slight reduction in the pressure level until automation again achieves the normal level. During these periods of slight pressure reductions, the structure does not quite possess its usual rigidity, but these periods are so short that the lessened rigidity is not very critical.

This stabilising characteristic under gust loading makes it permissible to design for maximum mean wind speeds, rather than for gust loading, mean minute speeds perhaps being the most appropriate. Meteorological measurements of wind speeds are generally taken at about a height of 10 m above the ground, and as many structures

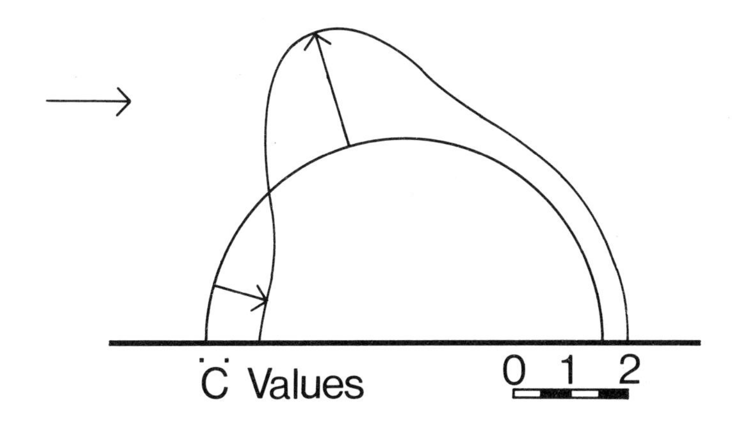

Fig. 34

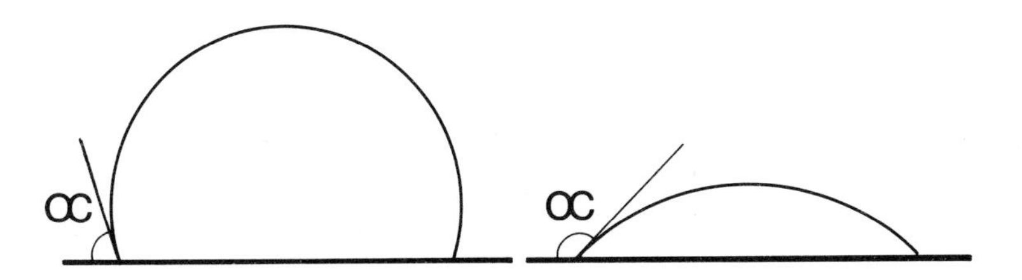

Fig. 35

* For a fixed mass of gas at a constant temperature the pressure is inversely proportional to the volume.

are below this elevation, an appropriate reduction in wind speed may be used. Berger and Macher have tentatively suggested a possible 20 per cent reduction,[11] but the British Building Research Station has made firmer suggestions for reduction coefficients which not only take into account building height but also consider the surrounding topography, and gust durations.[12] Due to its dynamic characteristics, the influence of wind loading on an air supported structure is much greater and more critical than with more conventional constructions and it is imperative that this loading influence is fully explored in the design of air supported structures. This is not to say that these structures are susceptible to wind loading; on the contrary there are numerous examples of pneumatics standing up to hurricane loading due to their resilience, whilst neighbouring more conventionally constructed buildings have suffered quite severe damage. However, failures have sometimes occurred in cases where the design of the pneumatic has not been given sufficient attention. Also, the magnitude of deformation common to a pneumatic can often severely hamper the building's function. It is for these reasons that the behaviour of air supported structures under wind loading must be more fully understood.

Rain Loading

In all normal circumstances rain does not cause a loading problem, since pneumatic forms allow rainwater to drain off the structure quickly without the formation of water pockets. With the simple spherical and cylindrical forms generally used at present for air supported structures there is certainly no possibility of rainwater collecting on the membrane. However, with flatter structures employing cable, net, indent and interior membrane wall construction there is some danger that water could accumulate in pockets during heavy rainstorms, causing deformation and subsequent collapse of the structure. In such cases careful consideration should be given to the provision of adequate drainage facilities.

Internal Air Pressurisation Loading

All these external surface loads are counteracted by the internal pressurisation loads, which must at all times and at all points on the membrane surface be greater or equal to the total sum of the external surface loads, so that no negative compressive stresses are induced in the membrane, as given by condition (1.1)

$$T_{min}\ `P_x + P_i\text{'} \geqslant 0$$

This, the folding condition, gives the required internal pressurisation for stabilisation under the various loading conditions. If this pressurisation is not sufficient to satisfy the above condition, local folding will occur, and under dynamic loading the membrane will flutter and

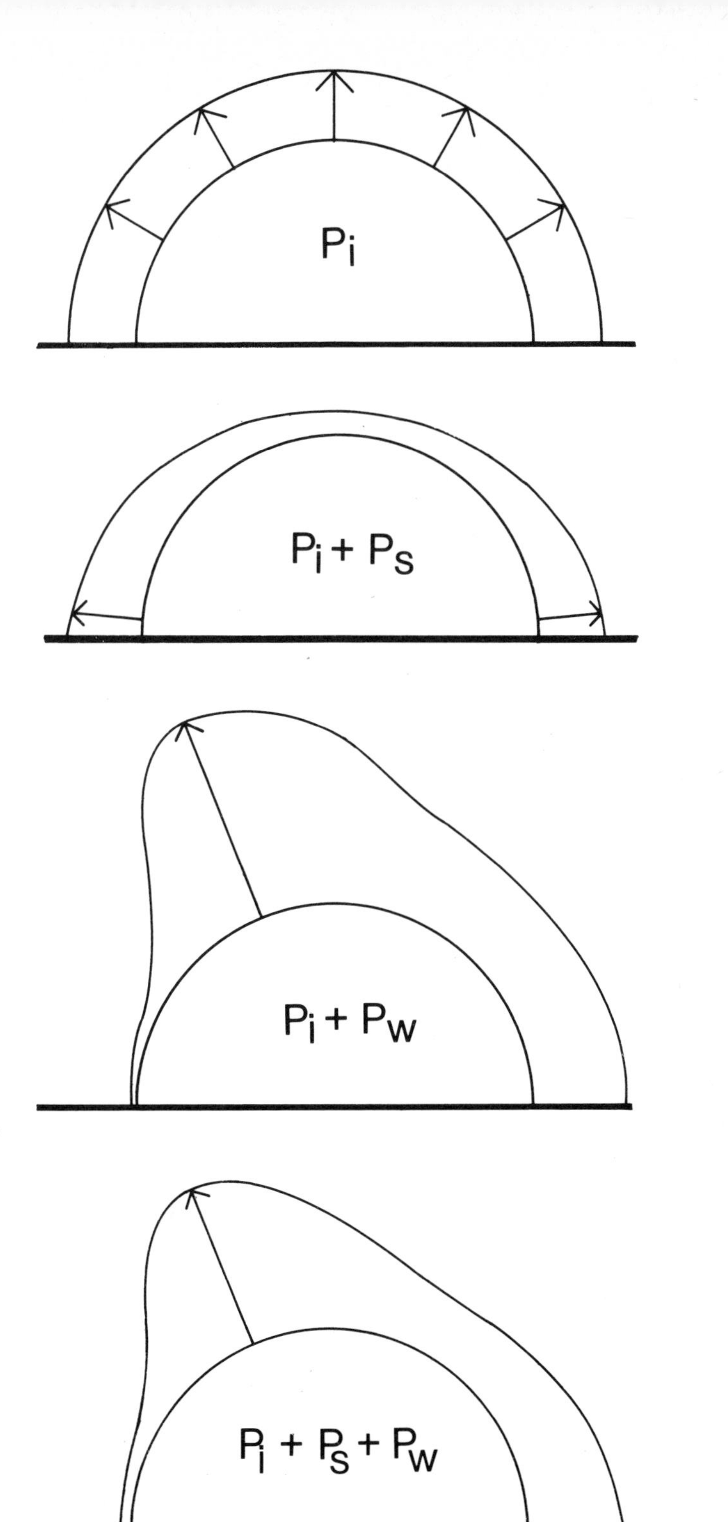

Fig. 36

Fig. 37

Fig. 38

Fig. 39

be subjected to shock stresses. These shock stresses severely reduce the structural life of the membrane material. It has been suggested by Ruhle that for less demanding uses, such as warehousing and sports, the folding condition criterion could be reduced, thus allowing some membrane flutter and larger deformations.[13] More searching investigations are essential before such reductions can be employed with confidence and certainly the contents of the air supported building would have to be kept some distance away from the membrane.

The dynamic character of the pneumatic is once again evident by the very fact that the internal pressurisation load can be altered to suit the varying external load conditions. However, this opportunity to maintain maximum structural efficiency at all times by automatic manipulation of the pressurisation element is in many cases not used to good advantage. In these cases the internal pressure is maintained at a constant level, sufficient to support the maximum external loading ever encountered. Having established the maximum internal pressurisation necessary to prevent folding under all loading conditions, it is essential to examine the maximum loading which occurs across the membrane, so that the stresses induced in the membrane can be predicted. This situation need not necessarily occur under the combined action of all the loadings, and therefore, all the different possible combinations of loading must be investigated. With simple pneumatic forms, without cable and net reinforcement, the worst condition generally occurs under the simultaneous action of the internal pressurisation loading and the wind loading, that is '$P_i + P_w$', as shown in the adjacent diagrams.

Fig. 36

Fig. 37

Fig. 38

Fig. 39

Pneumatic Membrane Stresses

In general, to predict the stresses induced in the pneumatic membrane by the various loading conditions, mathematicians employ the classic membrane equations applicable to shell structures, which are based on linear theory. Two major assumptions, both very dubious for pneumatics, are made. Firstly, that the membrane suffers only small deformation, so that loading can be considered as acting on the undeformed system, and secondly that for all practical purposes the stress–strain relationship of the materials can be treated as linear. Trostel uses even more complex analyses, which take into account the shear rigidity of the system.[14] Although this linear theory is quite valid for most conventional structures and materials its assumptions cannot be applied to pneumatic structures as was explained by Oden and Kubitza at the Stuttgart Pneumatic Colloquium, following their work on inflated rubber membranes[15]:

> 'The behaviour of inflatable pneumatic structures is inherently non-linear; such structures often acquire their primary load-carrying capacity after undergoing deformations which, even under small pressures, may be so large that the original undeformed shape is unrecognizable. Strains appreciably greater than unity are not uncommon, and in such cases Hooke's Law

> is not applicable. Moreover, the materials used to construct pneumatic structures are often anisotropic and non-linearly elastic and, to further complicate matters, the directions and magnitude of the applied loads change with the deformation. To emphasise this point, one need only refer to recent experimental studies on pneumatic structures wherein discrepancies of the order of 400 per cent were encountered when measured stresses were compared with those predicted by conventional linear theories.'

Consequently, there has been a trend towards the use of non-linear structural theories for pneumatic analysis, and this trend has been simplified by significant developments in both large scale digital computers and general methods for numerical analysis. However, these methods are rather tedious, and certainly, in view of the difficulty in mathematically representing loading conditions, they are at the moment rather unsuitable for design purposes. Consequently simple methods of analysis or even model analysis are more appropriate until the understanding of pneumatic behaviour is considerably more precise, and this can only be achieved by extensive research efforts.

Although it is extremely difficult to analyse the minimum surface forms achieved with soap films, expensive photographic analysis being about the only method, very similar minimum surface forms are produced when thin rubber membranes are inflated. By using grid-marked membranes it is very easy to measure the exact shape of the inflated form and determine the strain and stress magnitudes over the membrane surface. Such model analyses not only indicate the most suitable forms for pneumatics but can also be used to study the stress patterns generated in the membrane under different loading conditions. If these methods of analysis are combined with the following simple mathematical processes, then a prediction of pneumatic membrane stresses can be made that is reasonably accurate considering the variations of loading conditions that occur.

Fig. 40

The stresses at any point in a pneumatic membrane are determined by the total pressure differential across that membrane and by its curvature at that point and can be simply represented by:

$$T = f(P)f(R) \qquad (3.10)$$

At every point in this structural system, equilibrium must exist, and

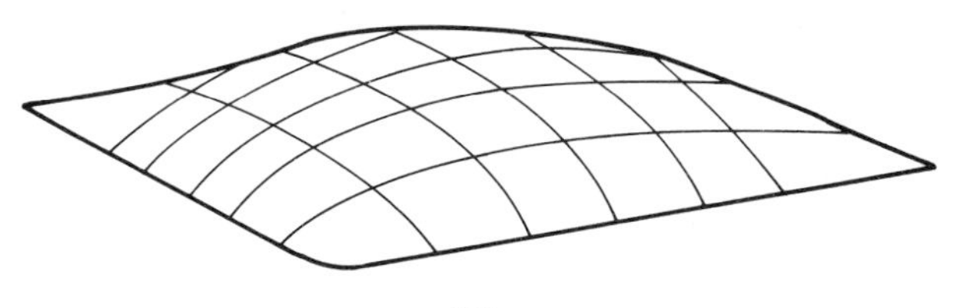

Fig. 40

these equilibrium conditions are maintained by deformation and subsequent redistribution of the stresses. For a spherical structure acted on by uniform pressure differentials, the stresses in the membrane are equal in both the principal planes, that is α equals β. If conditions vary, such that the stress in one direction cannot be fully reacted, then the difference in stress will be transferred to the other plane. A simplified prediction of the stresses conditions within a pneumatic is given by:

Fig. 41

Fig. 42

$$P=\frac{T_1}{R_1}+\frac{T_2}{R_2} \tag{3.11}$$

For a spherical form

$$R_1=R_2=R$$

$$\therefore T_1=T_2=T=\frac{P.R}{2} \tag{3.1}$$

For a cylindrical form

$$R_1=R \quad \text{and} \quad R_2=\infty$$

$$\therefore T_c=P.R \tag{3.12}$$

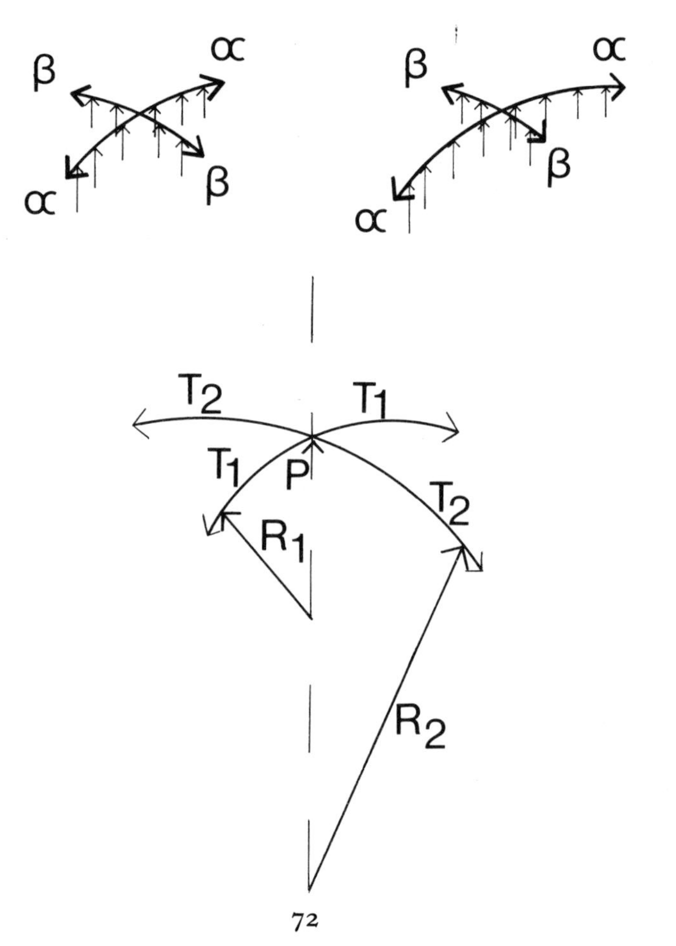

Fig. 41

Fig. 42

The longitudinal stress, 'T_1', is difficult to determine, being dependent on the end conditions. If the cylindrical membrane was fixed between rigid vertical end abutments, the internal pressurisation would be restricted by the abutments, and very little stress would be induced longitudinally into the membrane. If, however, the membrane were fixed between rigid end walls which were not anchored to the ground, internal pressurisation would induce longitudinal stresses in the membrane of the order predicted by the expression

$$T_1 = \frac{P.R}{2} \qquad (3.13)$$

The above expressions assume that the membrane is not deformed, so how then does deformation affect the validity of these expressions? When the membrane is deformed, the radius of curvature of the membrane at the point of maximum loading is reduced, and consequently the stress at this point will be less than that calculated, assuming the above expressions. This therefore provides an additional factor of safety. Perhaps graphical methods could be used to determine the deformed shape and consequently assess what reduction coefficients could be applied to the expressions so that a more accurate representation of stress conditions is indicated.

Fig. 43

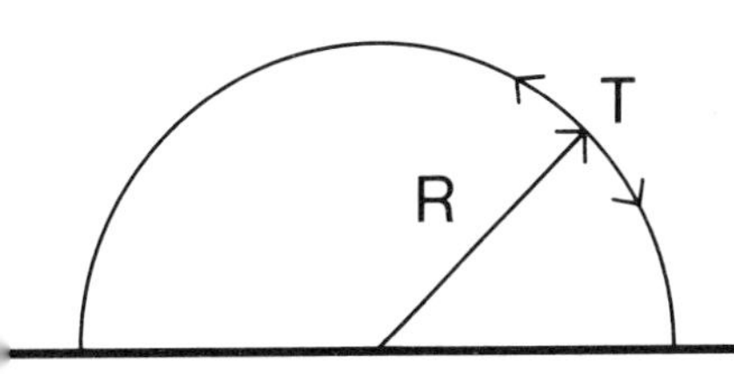

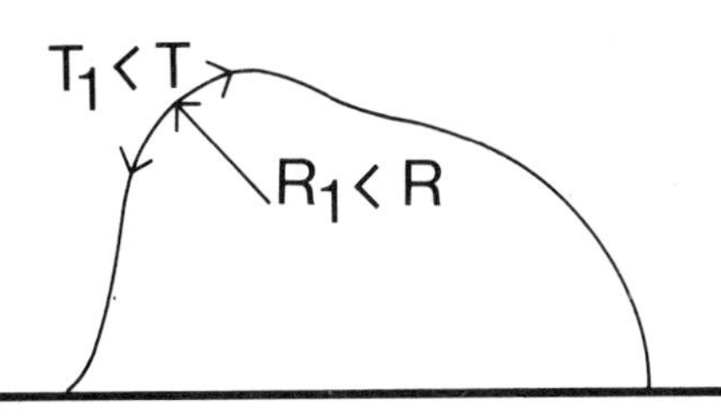

Fig. 43

Pneumatic Forms

Undoubtedly the sphere is the most perfect form for a pneumatic, the membrane being stressed equally at every point and in all directions. This form, the purest both structurally and aesthetically, is very rarely used, since functionally it is rather impractical. However, an air supported spherical structure, manufactured by Stromeyer, was used as an information pavilion at a trade fair in Berlin. The very first air supported radome structures were three-quarter spheres, and still this form is much utilised for the protection of radar equipment. However, these forms are not so easily fabricated, nor are they always practical, and so the common obsession of designers for rectilinear planning is also prevalent in pneumatics. For uses such as warehouses, factories, indoor sports and places of assembly, cylindrical forms on rectangular plans, with or without spherical ends have become most popular. Spherical ends to cylindrical forms are often used since the longitudinal membrane stress in the cylinder is the same as the stress in spherical end. However the

Plate 30

Plate 31

Plate 32

Plate 30 Air supported spherical structure used as an information centre for SPD at a trade fair in Berlin, manufactured by Stromeyer

Plate 31 An air supported structure of cylindrical form on a rectangular plan

Plate 32 An air supported structure of cylindrical form, with spherical ends

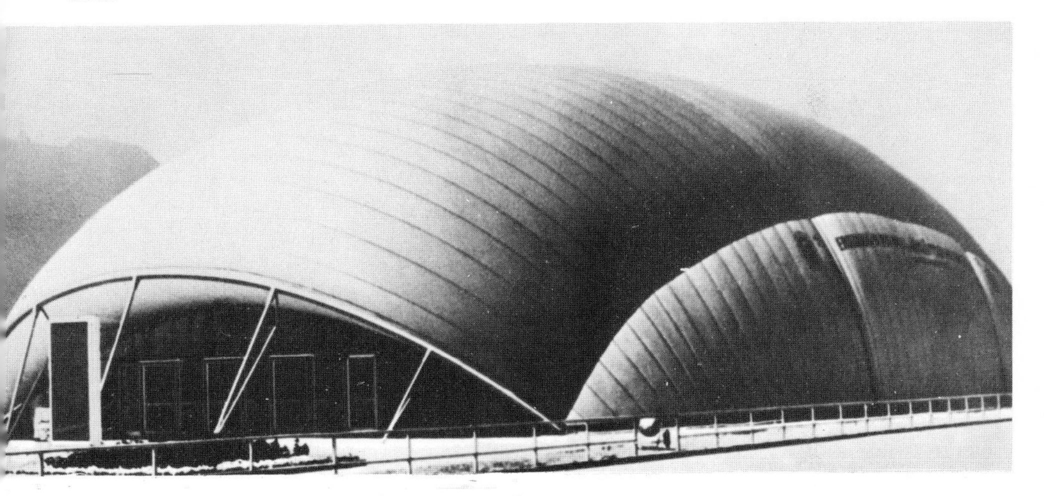

Plate 33 A torus shaped air supported structure

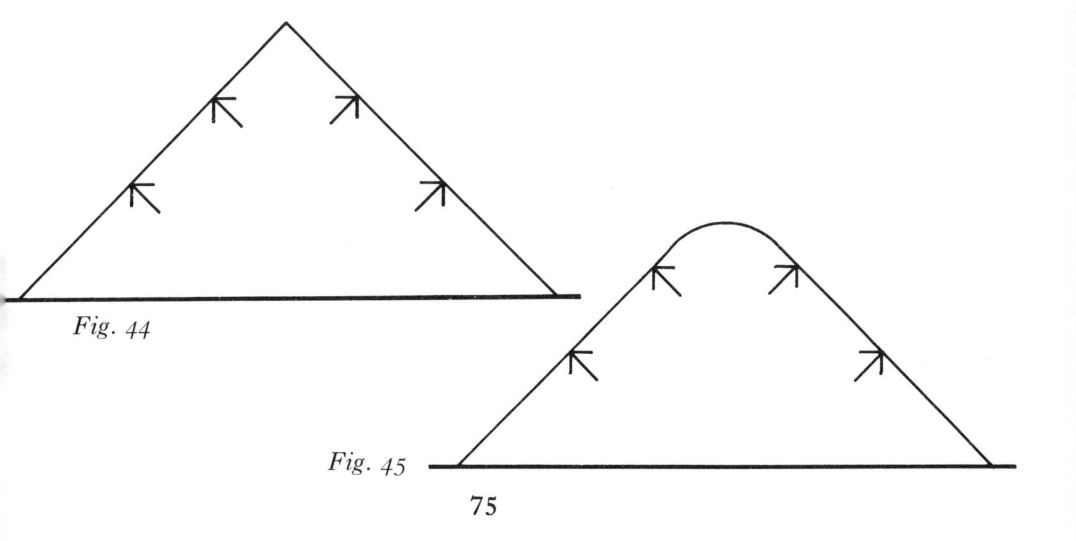

Fig. 44

Fig. 45

circumferential membrane stress in the cylinder is twice that of the sphere, resulting in distortion and local folding of the membrane at the junction between the two forms. A more harmonic transition for these end conditions can be achieved with elliptical or torus shaped sections, but such forms are extremely difficult to fabricate, especially when compared with cylindrical forms. On the other hand, the conical form lends itself ideally to fabrication. Unfortunately the pre-tensioning stresses are greatest at its base and decrease progressively as the peak is approached, with the result that the peak is very limp and can be considerably distorted by external loadings. The cone's rigidity can be increased by harmoniously rounding off the peak. Pneumatic conical towers of high stiffness have been successfully manufactured both in Great Britain and in America.

Plate 33

Fig. 44

Fig. 45

Except for a very few cases, designers have undoubtedly not explored the scope of pneumatics beyond these very simple forms. This is perhaps understandable since even with these forms knowledge of their behaviour is far from precise, but without exploration, imagination and a search for more sophisticated forms, pneumatic potential is likely to stagnate.

Cables, Cable Nets and Internal Membrane Walls

As the use of cables, cable nets and internal membrane walls become more and more prominent, so more imaginative designs will emerge. Cables and nets can divide the membrane into a number of small elements, with small radii of curvature, so reducing the membrane stresses. By deployment of nets, very thin membranes can be used; the membrane presses against the net, under the internal pressure, and the major portion of the stresses is transferred to the netting. The main forces can be taken up by cables, which form deep grooves in the membrane. In this way, the endless number of forms, suggested by soap bubble conglomeration, can be achieved pneumatically. Endless alternatives and combinations of plan shapes, with different cable and net arrangements, are possible. On the same plan area, the form of the pneumatic can be considerably altered by adjustment of the length and position of the cables. Circumferential cables are particularly suitable for high domed and tower forms, whereas for shallower forms radial cables are much more effective. Nets of uniform gauge are most efficient, since the surface is broken up into a number of equal sized elements, all of similar curvature. Netting, with a variety of mesh sizes, gives rise to widely different membrane stresses in the various elements, and consequently much of the membrane material will not be stressed to its maximum limit. Large spans can therefore be achieved with thin transparent membranes, by reinforcing them with a fine mesh netting, which in turn is supported by wide mesh nets and cables. With this utilisation of cables and nets, the basic pneumatic shapes can be transformed into an infinite number of spatial and sculptural forms and the designer is given much scope to

Plate 34

Fig. 46

Plate 34 Air supported transparent p.v.c. dome, reinforced by cable nets, designed by Arthur Quarmby for the 20th Century Fox film production, 'The Touchables'

use his imagination in moulding the structure to suit functional and aesthetic requirements. Although cable networks are already used in pneumatics to a limited extent, their exploitation falls far short of the bold and imaginative forms suggested by Frei Otto.[16] The U.S.A. Pavilion at EXPO '70 is a rare exception and certainly lends hope to the future.

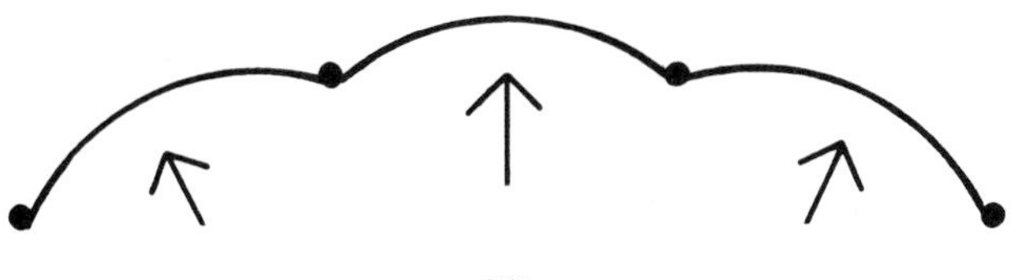

Fig. 46

Only a few mathematicians have concentrated their efforts on this field, notably Cowan and Gero, who have proposed a method of analysis using an iteration technique suitable for digital computer programming.[17] The membrane presses against the cable network, and consequently, the loading of the cable is a function of the pressure differential, which acts normally to the cable and is transmitted to it by the membrane stresses, and the deadweight of the cables, acting vertically downwards. The analysis of these loading conditions is extremely complex, and would hardly be achieved without the aid of computer techniques.

Boundary Membrane Stresses

Up to this point, the discussion of stress conditions within the pneumatic membrane has disregarded the effect of boundary restrictions, such as anchorages and openings in the membranes, but these restrictions can influence the stress conditions quite substantially. Equilibrium conditions within a pneumatic membrane are maintained by deformation of the membrane and subsequent redistribution of the stresses throughout the membrane. Boundary restrictions reduce this deformation causing local stressing of the membrane within the vicinity of these boundaries. With normal anchorages in a single horizontal plane the stresses induced due to the restricting influence of the fixing, can usually be taken up by the membrane. This is because the loading across the membrane in these areas is generally small, causing the membrane to be only lightly stressed and leaving it capable of accommodating quite significant stress increases due to boundary restrictions. On the other hand, when rigid elements, such as doorways, are attached to the membrane, severe local stress concentrations can build up under high loading, since redistribution of the stresses is not possible through deformation. In these circumstances, either the membrane must be reinforced, or a flexible junction be provided between the rigid element and the membrane so that the latter is free to move allowing stress redistribution. Since an exact prediction of stress conditions is extremely difficult, the latter method for the design of membrane openings is more prudent. However, if the former method is used, valuable stiffening of the whole structure is accomplished.

Anchorage Loads

Unlike conventional structures, which exert a positive loading on the ground, the pressure differential across the membrane of an air supported structure causes uplift forces on the whole of the structure, and these forces must be resisted by firmly anchoring the structure to the ground. In the early days of pneumatic building construction a few structural failures occurred because of lack of sufficient anchorage provisions. In these instances anchorage design was by trial and error and was not based on any theoretical calculations whatsoever. Through these trial and error methods anchorage designs have been evolved, which are quite satisfactory for typical air supported forms. However, if pneumatics are to become universally accepted as a serious form of building construction, it is essential that the extent and nature of these uplift forces be known so that suitable anchorage provision can be more precisely designed. The total vertical uplift force is equal to the sum of the maximum aerodynamic lift and the maximum load due to the inflation pressure, and this is resisted uniformly by the base anchorage. For a simplified assessment of the magnitude of the anchorage reaction, the loading of the membrane can be represented as having a uniform average value throughout. This average value can be taken as acting on the horizontal projection of the pneumatic form and therefore the total vertical uplift force is the product of the average membrane loading and the plan area of the structure. The force is uniformly reacted around the structure's perimeter and consequently the vertical anchorage reaction per unit length is given by the following expression.

$$V = \frac{P_a . A}{L_p} \tag{3.14}$$

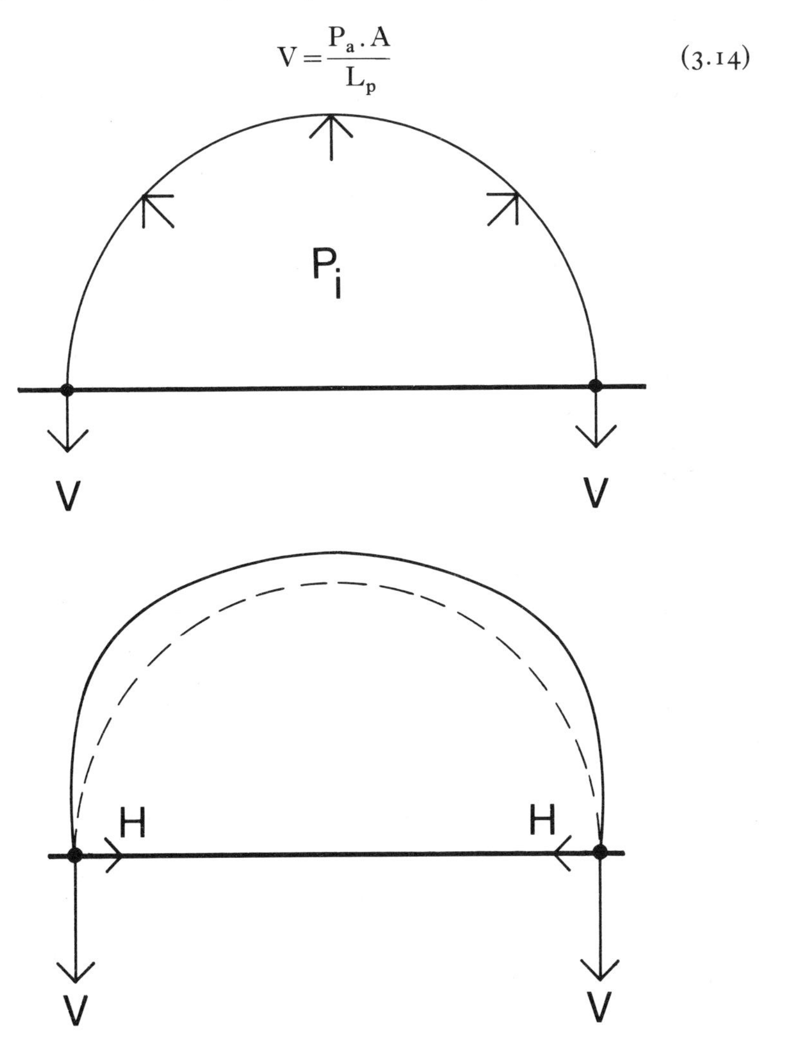

Fig. 47

Fig. 48

With forms which possess a base angle of 90°, such as a semi-cylinder and a hemisphere, this vertical anchorage reaction is equal to the membrane tension at the boundary. If it is assumed that such forms are not deformed then there will be no horizontal reaction at the anchorage. However, this assumption is not quite valid because of the boundary conditions which have already been mentioned. Due to the pressure differential loading across the membrane, the membrane will extend and the structural form will increase in size as indicated. Since the anchorage prevents the movement of the membrane, small horizontal reactions exist which increase in magnitude as the membrane loading increases. For all practical purposes, though, the horizontal anchorage reactions of these forms are so small that they can be ignored, since whatever form of anchorage is used, the very act of anchoring the membrane provides quite a degree of resistance against horizontal forces.

Fig. 47

Fig. 48

Where the base angle of the pneumatic form is not a right-angle there is a horizontal as well as a vertical reaction at the base, the actual magnitude of these reactions being a function of the base angle as indicated by

Fig. 49

$$V = T . \sin \alpha \tag{3.15}$$

Fig. 50

$$H = T . \cos \alpha \tag{3.16}$$

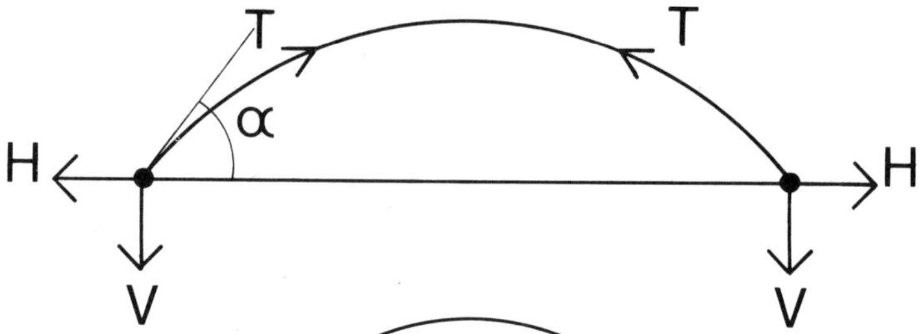

T
T
α
H
H
V
V

Fig. 49

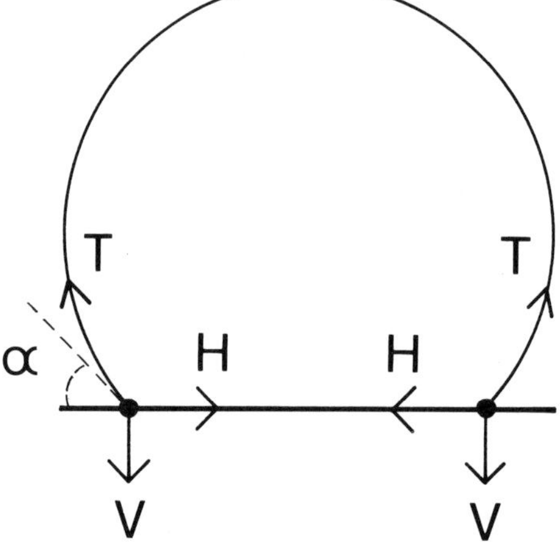

T
T
α
H
H
V
V

Fig. 50

The above expressions are undoubtedly an over simplified statement of boundary conditions. The exact magnitude of boundary membrane tensions is extremely difficult to predict, so these expressions employ the maximum membrane stresses which occur at positions of peak loading, and these generally do not occur in the boundary regions. Moreover, horizontal hoop tensions in the membrane provide a partial horizontal reaction thus reducing slightly the magnitude of the horizontal reaction that the base anchorage must account for. Consequently these expressions provide a built-in safety factor for anchorage design.

With cable and cable net structures anchorage forces are concentrated at points rather than distributed uniformly around the perimeter base. In large structures these point forces are very large and substantial anchorages have to be provided, as was certainly evident in the U.S.A. Pavilion at EXPO '70.

Inflation Equipment Design

In the design of pneumatic structures, the structural engineer is presented with a problem totally foreign to his training and experience. To ensure structural stability, inflation equipment must be designed, and for this task a ventilation engineer is better qualified. As with the anchorage system this inflation equipment has on the whole been designed by trial and error. Calculation on air losses are extremely difficult and cannot hope to be accurate, since these losses are dependent a great deal on how much traffic movement, pedestrian or otherwise, there is in and out of the building. The capacity of the inflation equipment is also dependent on the required environmental standards and thus it is essential that its design be the result of the integrated efforts of structural and environmental engineer. The detailed design of inflation equipment is examined in the following chapter, because of the many practical factors that must be considered.

Design Safety Factors

Whenever a new structural system is adopted, whose behaviour is not thoroughly understood, it is essential that the design safety factors employed provide more than adequate safety. The use of pneumatics is so diversified that a general safety factor would only be appropriate in a few instances and so a number of different safety factors are necessary so that the whole range of pneumatic use is adequately catered for. The choice of a suitable safety factor is influenced by the ageing properties of the material used and the functions for which the structure is designed. With the fabric coated materials at present used for pneumatic construction, there may be substantial losses of material strength due to ageing. These losses are dependent to quite an extent on the nature of the coating which protects the base fabric. The overall safety factor must accommodate both these influences, and ideally should be divided into two components, one which takes account of the material strength loss and the other which caters for the building function.

$$\text{Total safety factor} = \frac{\text{Max. limit stress of material}}{\text{Usable design stress}}$$

$$Y = \frac{T_m}{T_p} \tag{3.17}$$

where $$Y = Y_1 . Y_2 \tag{3.18}$$

Throughout the world the value of 'Y_1', the material strength loss safety component, varies between 2 and 4, though the tendency is certainly towards the higher value. As far as 'Y_2', the function safety component, is concerned, its value could perhaps vary between the extremities of 1 and 2, the lower extremity being applicable to short-term enclosures for storage and the upper level being suitable for public buildings demanding high standards of safety. This component could perhaps be stipulated for anchorage design.

To conclude this analysis of the behaviour of air supported structures, an example will aid the clarification of the design methods proposed in this chapter. The design to be considered is for an exhibition pavilion of three-quarter spherical form for use in the United Kingdom for a period of up to 10 years. The pavilion has a plan diameter of 30 m and its spherical radius is 22 m. A typical wind loading value in the UK can be taken as 45 m/sec. However, British regulations permit this figure to be reduced to 39·5 m/sec owing to the short life of the building.

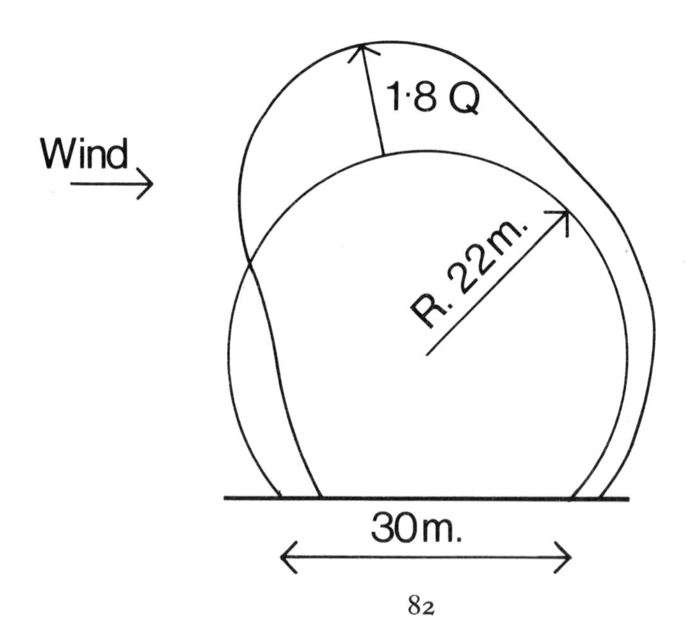

Fig. 51

Maximum wind speed loading, $v = 39{\cdot}5$ m/sec
Dynamic pressure of wind, $Q = 950$ N/m^2

Folding condition for three-quarter sphere:

$$\frac{P_i}{W} \approx 1$$

Therefore maximum internal pressure,

$$P_i = 950 \text{ N/m}^2$$

$$P_i \equiv 97 \text{ mm of water}$$

It is essential to check the folding condition under the maximum snow loading condition. In the United Kingdom a snow loading of 75 kgf/m^2 is usual. The maximum snow loading as given by the expression (3.2) occurs when 'θ' is a right angle.

$$\therefore \; P_s = 75 \text{ kgf/m}^2$$

$$P_s \equiv 75 \text{ mm of water}$$

Since 'P_i' is considerably greater than 'P_s' folding will not occur under snow loading. *Fig. 51*

Maximum wind pressure loading,

$$P_w = 1{\cdot}8Q$$
$$= 175 \text{ kgf/m}^2$$

Maximum pressure loading on membrane

$$P = P_w + P_i$$
$$= 175 + 97 \text{ kgf/m}^2$$
$$= 272 \text{ kgf/m}^2$$

Tensile stress in membrane,

$$T_m = \frac{P.R}{2}$$

$$= \frac{272 \times 22}{2} \text{ kgf/m}$$

$$= 2992 \text{ kgf/m say } 3000 \text{ kgf/m}$$

$$T_m \equiv 150 \text{ kgf/50 mm strip of material}$$

Since the building is likely to be erected and dismantled many times during its life span the membrane will be subject to considerable abuse and so a safety factor component for material strength loss of 3 is advisable. As far as the function safety component is concerned, the building, although only short term, will be used by the general public, and consequently a safety factor of 1·33 is recommended. This gives an overall safety factor of 4. A membrane material of tensile strip strength 600 kgf/50 mm strip is therefore needed. Because of the abuse the material is likely to suffer a high tearing strength is recommended. A suitable material would perhaps be a polyester fabric coated with polyvinyl chloride of overall weight about 1 kg/m^2.

Since the base angle of the three-quarter sphere is 45°, the horizontal and vertical components of the anchorage reactions will be equal, i.e.

Fig. 52

$$H = V$$

The average wind loading over all the membrane can be taken as 0·75Q, thus the average pressure loading across the membrane is given by:

$$Pa = (0{\cdot}75).(97) + (97)\ \text{kgf/m}^2$$
$$= 170\ \text{kgf/m}^2$$

Now

$$V = \frac{P_a . A}{L_p}$$

$$= \frac{170\pi . 15^2}{2\pi . 15}\ \text{kgf/m}$$

$$= 1310\ \text{kgf/m}$$

Anchorage reactions V = H = 1310 kgf/m length of perimeter.

Assuming that a safety factor of 1·33 is used for the anchorage design, the type of anchorage used must be designed to resist a horizontal and vertical reaction of 1750 kgf/m length of the structure's perimeter.

This example gives some idea of the magnitude of membrane stresses and the anchorage forces that occur in a typical pneumatic structure. Although the above predicted magnitudes are by no means precise, their accuracy compares favourably with predictions by so-called exact analytical theories, and certainly this simple design

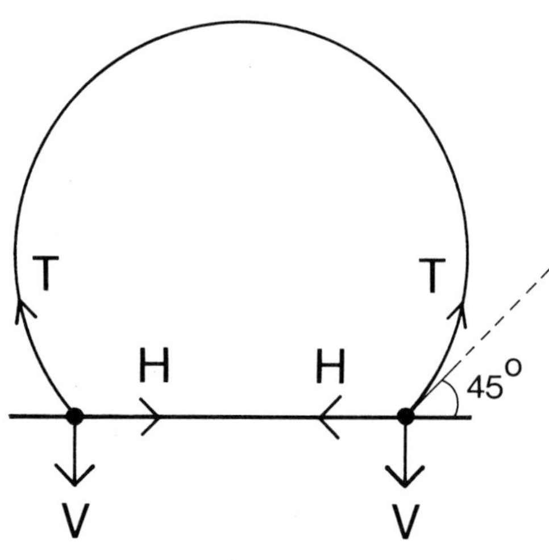

Fig. 52

method seems very appropriate in these early days of pneumatic building construction.

AIR INFLATED CONSTRUCTION

Unlike air supported construction, this form of construction is much closer to more conventional constructional techniques in its behaviour. Inflated members form conventional structural elements, such as beams, columns, arches and walls, which resist the loading conditions associated with conventional rigid structures, although these can fluctuate slightly due to deformations. Much research has been undertaken into the structural properties of air inflated members for use in space, particularly by the U.S. National Aeronautics and Space Administration. This research has been concerned mainly with inflated cylinders and 'airmat' construction. However, this scientific and intensive approach to air inflated construction does not extend to the building industry; here pneumatic design is usually by trial and error.

Factors Influencing the Design of Air Inflated Elements

Four factors influence the type and magnitude of loading that an inflated member can carry; these are the volume of air contained within the inflated member, the internal air pressure, the structural form of the member, and the characteristics of the membrane material. A simplified analysis of an inflated cylindrical beam under bending is as follows.
Pretensioning stresses due to internal pressure

$$T_1 = \frac{P_i . R}{2}$$

Fig. 53

$$T_c = P_i . R$$

For a uniform distributed load 'W' on a simply supported beam the maximum bending moment is:

$$M = \frac{W . L^2}{8} \quad \text{or} \quad M = W . f(L^2)$$

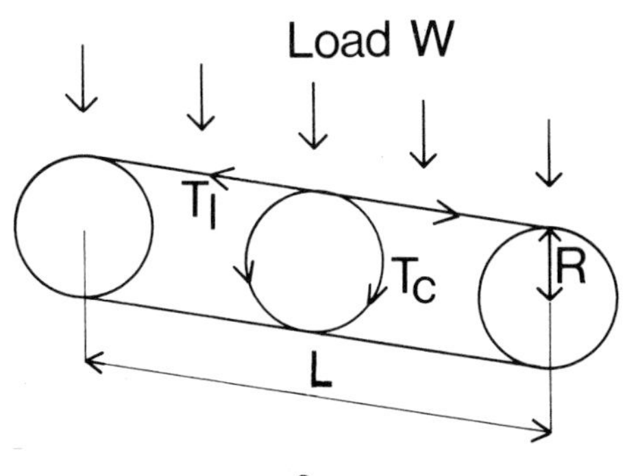

Fig. 53

$$\text{Stress due to this bending moment, } T_w = \pm\frac{M}{Z}$$

Where 'Z' the section modulus of the beam is a function of the cube of the radius of the beam

$$\therefore\ T_w = W.\frac{f(L^2)}{f(R^3)} \qquad (3.19)$$

Fig. 54

To prevent folding of the membrane material the pretensioning stresses in the upper fibres must not be exceeded by the compression stresses due to bending as indicated by

$$T_1 \geqslant T_w \qquad (3.20)$$

$$\frac{P_i.R}{2} \geqslant W\cdot\frac{f(L^2)}{f(R^3)}$$

$$P_i \geqslant W\left[\frac{2}{R}.\frac{f(L^2)}{f(R^3)}\right]$$

Since for a given beam, the length and radius are a constant, the above expression indicates that the external loading an air inflated member can carry is proportional to the internal air pressure.

$$P_i \geqslant k.W \qquad (3.21)$$

However, the regions of zero stress at the folding conditions are very localised, so the beam can in fact take greater loading before collapse occurs. As the loading is increased so the folds in the membrane spread round the section, until at the collapse point the region of zero stress extends around the whole section. At this point collapse in the true meaning of the word may not necessarily occur, in other words the beam recovers its form on removal of the load, provided that the membrane has not failed. If loading does continue beyond this point the lower membrane fibres will act as a suspension structure. The collapse load is found to be approximately double the

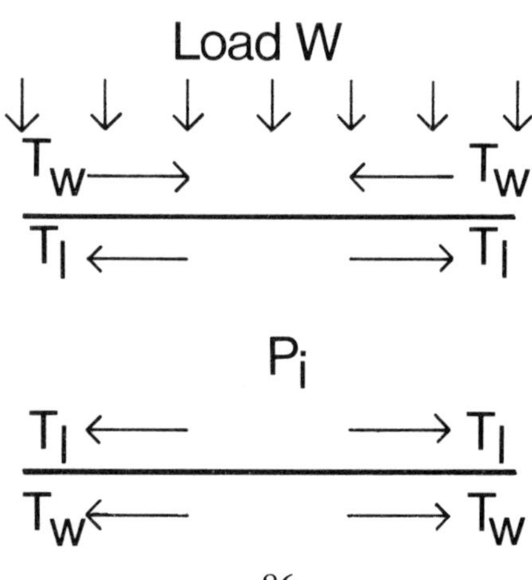

Fig. 54

wrinkling or folding load. At the point of folding it can be seen that the maximum stress in the membrane is

$$T = T_l + T_w \tag{3.22}$$

This maximum stress must not exceed the usable design stress of the membrane material

$$T_p \geqslant T_l + T_w \tag{3.23}$$

Fig. 55

$$T_p \geqslant \frac{P_i . R}{2} + W \cdot \frac{f(L^2)}{f(R^3)}$$

$$P_i \leqslant \frac{2T_p}{R} - W\left[\frac{2}{R} \cdot \frac{f(L^2)}{f(R^3)}\right]$$

$$P_i \leqslant \frac{2T_p}{R} - k . W$$

Thus the limiting values for the internal air pressure can be represented by the following expression, where 'k' is a constant that describes the beam's dimensions.

$$k . W \leqslant P_i \leqslant \frac{2T_p}{R} - k . W \tag{3.24}$$

The above analysis assumes that the beam is subject only to bending stresses and ignores the effect of shear stresses. Shear stresses can produce folding of the membrane even though the longitudinal and circumferential stresses are both tensile. This is particularly so when the beam is subject to point loading. In the neighbourhood of the loading points high shear stresses can cause folding of the membrane on the sides of the beam, following which premature collapse of the beam can occur. If failure of the beam is due to shear, the actual collapse loads are generally only slightly higher than the folding loads. This behaviour of the beam under shear is completely different from failure that is caused by bending stresses. It is therefore important that both shear and bending are considered in the design of air inflated elements.

The other major design criterion that must be considered is the deflection or stiffness of the member. Unfortunately theoretical work

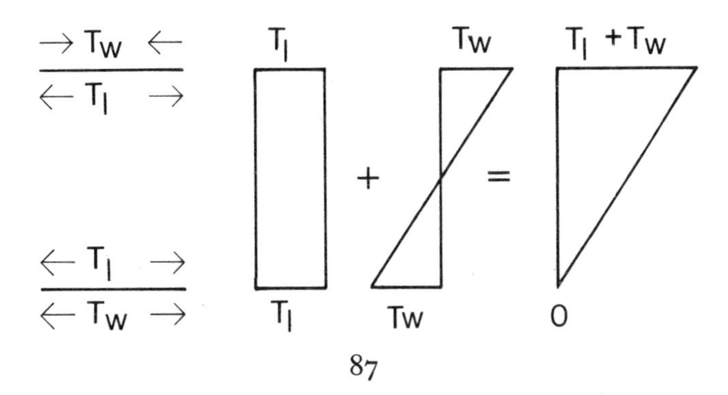

Fig. 55

Fig. 56 on this is not very clear, although certain generalisations can be made. Load–deflection relationships appear to be non-linear, as does the relationship between internal air pressure and deflection for a given *Fig. 57* load. The adjacent diagrams illustrate the sort of relationships that can be expected. All fabric materials have marked creep tendencies, that is deflection for a given load increases with time and consequently load–deflection curves exhibit a hysteresis effect.

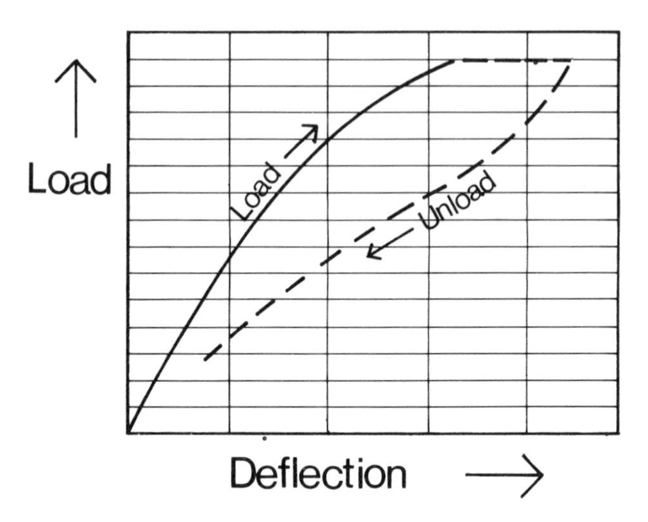

Fig. 56

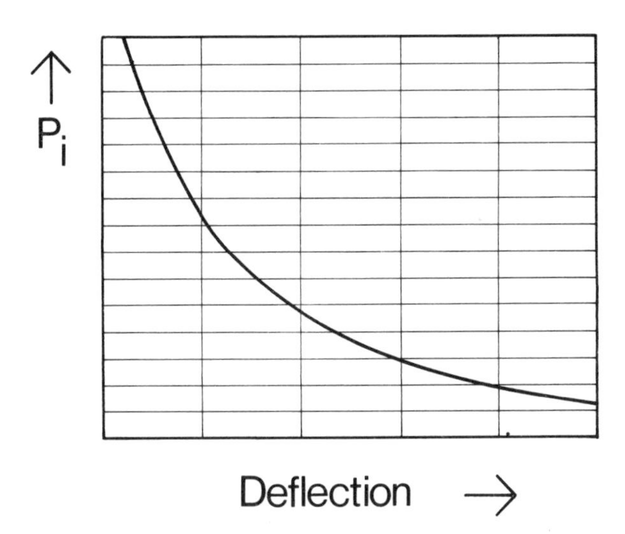

Fig. 57

Summary of Structural Behaviour of Air Inflated Elements

The above observations have been concerned in the main with a simple cylindrical air inflated beam. Analysis of the structural behaviour of more complex air inflated forms, such as the airmat, is much more difficult. However, the following generalisations can be made for air inflated members.

1 The load carrying capacity is directly proportional to the internal pressurisation.
2 As the span of a member increases so its load carrying capacity decreases; this relationship is non-linear.
3 The load carrying capacity increases with the cross-sectional area of the member; in the case of dual walled elements, such as the 'airmat', it increases as the distance between the two membranes increases, in other words the load carrying capacity increases as the section modulus increases.
4 Stronger and stiffer membrane materials increase both the load carrying capacity and the rigidity of a member.
5 The stiffness of the member and its carrying capacity is greatly influenced by both the form of the member and its internal construction. Arched and domed forms, as is usual, have greater strength than straight beams and flat plates. Internal diaphragm or tie construction also adds to the strength of the member, diagonal connections being more effective than straight direct ones like the drop thread in 'airmat' construction.

Because of their efficient material deployment, air inflated structures are very lightweight and this means that the aerodynamic lift-off forces of the wind are not sufficiently counterbalanced by the structural mass. Positive anchorage of the structure to the ground is therefore essential. Although these anchorage forces are not of the magnitude of those associated with air supported construction, similar designs for the anchorage system can be used. The safety factors discussed previously are also appropriate for air inflated construction, although for high pressure structures even higher safety factors might be considered, since membrane material failure in these structures cause sudden release of the pressurised air which can be extremely dangerous. In these cases of failure the structure does not collapse, but explodes, and certainly such failures are considerably more disastrous.

HYBRID CONSTRUCTION

The term 'hybrid' describes a very wide range of constructional techniques. A wholly pneumatic hybrid utilises both air supported and air inflated construction, but, besides this, hybrid construction defines any structural technique in which pneumatic construction is combined with other forms of construction. So broad is the range of hybrid structures that examination of the behaviour of all the various types is practically impossible. In addition to this the constructional complexities of these structures is far greater than either air supported or air inflated structures, neither of which is easy to analyse precisely, as has been previously pointed out. So few hybrid structures have been designed and built that it is difficult to evaluate theoretical analyses wherever they have been attempted. Specific examples are discussed elsewhere in this book, showing how hybrid construction has contributed to their structural performance.

THE WAY AHEAD

As with any newly adopted constructional system, a full and precise analysis of pneumatic behaviour has yet to be achieved. Certainly great steps forward would be made if communication between pneumatic designers was improved. In many cases research is isolated around specific and very specialised problems, which advance pneumatic knowledge in a very narrow field, but contribute little to a general understanding of pneumatic behaviour. This chapter has reviewed the analytical knowledge of this behaviour and has discussed which theoretical methods are at present most appropriate for the design of pneumatic structures. These design methods can be summarised as follows:

1 Choice of suitable pneumatic forms using
   a Soap films and bubbles
   b Frei Otto's laws of formation
   c Model analysis
2 Structural design considering
   a Loadings

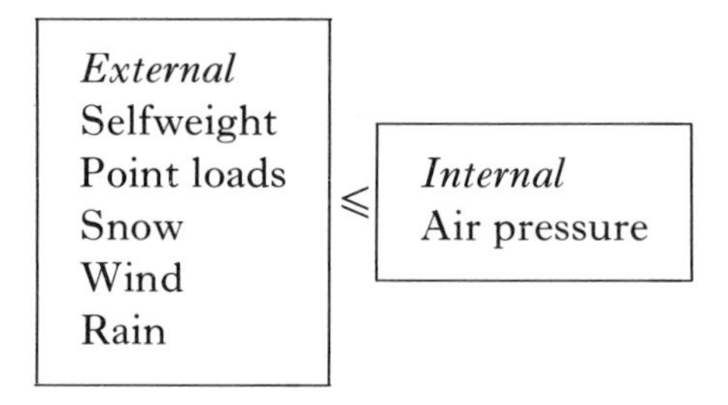

   b Membrane stresses—use of cables, cable nets and internal membrane walls
   c Boundary conditions
   d Anchorages
   e Inflation equipment
3 Choice of suitable safety factors.
4 Choice of suitable materials and equipment.

This methodical approach to pneumatic design is appropriate whether the mathematical design is simple, as suggested here, or is extremely complex and more accurate, as will be probable in future years. This discussion besides presenting a simplified explanation of pneumatic behaviour, has also revealed where knowledge is lacking, in the hope that it will act as a catalyst and stimulate pneumatic designers to search for a fuller more precise understanding of this new structural building form.

1. C. V. Boys, *Soap Bubbles, their formation and the forces that mould them.*
2. F. Otto, *Zugbeanspruchte Konstruktionen*, Band 1, p. 11 ff.
3. F. Otto, op. cit., p. 18 ff.
4. *Pneumatic Building Construction*, the Academy of Building and Architecture, U.S.S.R., p. 120.
5. H. Ruhle, 'Development of Design and Construction in Pneumatic Structures', *Proceedings of the 1st International Colloquium on Pneumatic Structures*, University of Stuttgart, 1967, p. 26.
6. G. Röntsch and F. Böhme, 'Model Analysis of a Semi-Cylindrical Air Supported Hull', *Proceedings of the 1st International Colloquium on Pneumatic Structures*, University of Stuttgart, 1967, p. 147 ff.
7. M. Kamrass, *Wind Tunnel Tests of a 1/24 Scale Model Air Supported Radome and Tower*, C.A.L. Report No. UB-909-D-1.
8. G. Berger and E. Macher, 'Results of Wind Tunnel Tests on Some Pneumatic Structures', *Proceedings of the 1st International Colloquium on Pneumatic Structures*, University of Stuttgart, 1967, p. 142 ff.
9. See *Proceedings of the 1st International Colloquium on Pneumatic Structures*, University of Stuttgart, 1967, p. 146.
10. See F. Rudolf, 'A Contribution to the Design of Air Supported Structures', *Proceedings of the 1st International Colloquium on Pneumatic Structures*, University of Stuttgart, 1967, p. 129.
11. G. Berger and E. Macher, op. cit., p. 145.
12. Building Research Station Digest 119, *The Assessment of Wind Loads*, Tables 1 and 2, H.M.S.O. publication, July 1970.
13. H. Ruhle, op. cit., p. 26.
14. R. Trostel has documented his analyses in F. Otto, *Zugbeanspruchte Construktionen*, Band 1, p. 170ff, and in a paper 'On the Analysis of Membranes', *Proceedings of the 1st International Colloquium on Pneumatic Structure*, University of Stuttgart, 1967, p. 78 ff.
15. J. T. Oden and W. K. Kubitza, 'Numerical Analysis of Pneumatic Structures', *Proceedings of the 1st International Colloquium on Pneumatic Structures*, University of Stuttgart, 1967, p. 87.
16. F. Otto, *Zugbeanspruchte Konstruktionen*, Band 1.
17. H. J. Cowan and J. S. Gero, 'Pneumatic Structures constrained by networks', *Proceedings of the 1st International Colloquium on Pneumatic Structures*, University of Stuttgart, 1967, p. 134 ff.

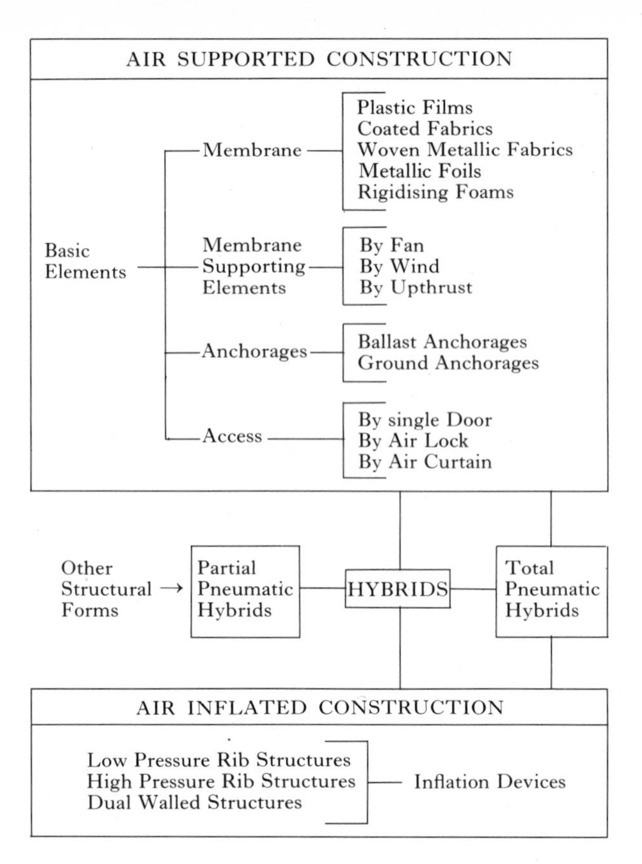

Fig. 58

4. Techniques of Pneumatic Construction

The above diagram illustrates the relationship between the various types of pneumatically stabilised building construction and how each type breaks down into a number of basic elements. Each of these are discussed in detail in the following pages along with the problems that are unique to this form of construction, such as environmental, the erection and dismantling procedure, cost, and also the extent to which the Building Regulations control its standards.

AIR SUPPORTED CONSTRUCTION

Basic Elements of Air Supported Construction

This type of construction possesses four basic essential elements, the structural containing membrane, the means of supporting this membrane, the means of anchoring it to the ground, and the means of access in and out of this environmental container. Membrane technology follows much the same lines as other forms of tensioned structures; inflation technology draws the structural engineer into the province of mechanical engineering; anchorage problems have already been encountered in suspension structures; consequently the design of these three former elements can derive much from already familiar techniques. However the provision of access against an air pressure differential presents the designer with a new problem, its only vaguely similar predecessor being in submarine engineering. Poor design of any one of these elements could lead to structural collapse and so each must receive equal attention if safe methods of construction are to be attained.

The Structural Membrane : Design Criteria

The choice of membrane material is dependent on a great number of design criteria, which can vary considerably with the function and design life of the structure. Before choosing a material the designer must evaluate these criteria with respect to the particular structure he is concerned with. The following criteria are suggested:

1 Strength
  a Tensile strength
  b Tear resistance
  c Modulus of elasticity
  d Strength to weight ratio
2 Air permeability
3 Weather resistance properties
  a Precipitation
  b Wind
  c Ultra-violet rays
  d Resistance to abrasion
  e Retention of physical properties with ageing
  f Atmospheric pollution
4 Chemical resistance
5 Insulation properties—transmission of heat, sound and light
6 Flexibility
7 Incombustibility
8 Variation of properties—especially with climatic changes
9 Strength of jointing methods—there is little point in having a high strength material if it cannot be jointed in a manner which takes full advantage of this strength.

At the present time four basic types of material are employed for pneumatic construction, plastic films, coated fabrics, woven metallic fabrics and metallic foils. Although, at the moment, coated fabrics are most popular, it is probable, in the future, that the other types will assume more significance than they do at present.

Plastic Film Membranes

Many designers are obsessed with transparency, which enables appreciation of man's total environment, without the accompanying inflictions of natural climatic conditions and human manufactured pollution; this is the attraction of transparent plastic films. Unfortunately, the films until recently available, did not have sufficient strength or satisfactory weathering properties. Transparent films are very susceptible to ultra-violet deterioration, but these harmful rays can be filtered out by the addition of suitable chemicals, such as the derivatives of benzophenon and benztriazol. Such films as polythene, polyethylene and unsupported polyvinylchloride have not the required strength for larger structures, unless used in conjunction with cable nets, and besides this, these films deteriorate very rapidly. However, recently developed materials, such as polytetrafluorethylene, are light, strong and durable, and very suitable for pneumatic structures. As yet such films are not much utilised, but no doubt once their suitability has been established, their use will be widespread.

Only one method of jointing is suitable for plastic films, and that is welding. Both sewing and cementing reduce the strength of the film considerably. With welding, 90 per cent of the material's strength can generally be developed, and in some cases, joints, as strong as the material itself, can be produced.

The table below gives the properties of various transparent films.

Material	**A**	**B**	**C**	**D**	**E**
*Hostavan**	West Germany	0·04	53	45	36
*Mylar**	U.S.A.	0·015	16	8·2	5·5
Polyethylene	U.S.S.R.	0·065	53	4·5	4·5
P.V.C.	U.K.	0·25	285	36	32

* Mylar and Hostavan are trade names for polytetrafluorethylene.

A Country of manufacture
B Thickness, mm
C Weight, g/m^2
D Longitudinal strength, kgf/50 mm width
E Transverse strength, kgf/50 mm width

Coated Fabric Membranes

At the present time, pneumatic structures are almost exclusively fabricated from synthetic fabrics, such as the polyamide fabric nylon and the polyester fabrics terylene and dacron, which are coated on one or both sides with vinyl, butyl, neoprene, hypalon or any of the other many plasticised elastomers. Although the coatings cause only very small losses in the strength of the base fabric, principally a reduction in tear strength, they considerably lengthen the time during which the fabric retains its mechanical properties under atmospheric exposure.

This reduction in tear strength, which is not very critical, is due to the fibres being twisted during weaving. The coating locks in this twist, consequently causing stress conditions within the fabric.

As far as the corroding influences of the atmosphere are concerned, the ultra-violet rays of the sun cause most deterioration. This deterioration is most marked with materials possessing transparent and translucent characteristics, and consequently where these properties are required, special measures must be taken. This can be achieved by using a base fabric, such as vinyl glass, which does not suffer such severe deterioration, or by filtering out these harmful rays by addition of suitable chemicals to the coating material to prolong service life. Other influences include temperature, humidity, precipitation, oxidisation and the presence of corroding gases in the atmosphere.

Coated Fabrics : The Base Fabrics

In nearly all cases synthetic materials are used for the base fabric although natural fabrics were employed in Russia for early structures. The high strength–weight ratio of synthetic fabrics is universally known, and indeed polyester fibres are among the best in this respect. Ageing characteristics are much superior with synthetic fabrics, especially polyester fibres which are even less susceptible to ultraviolet deterioration. The make-up of synthetic fabrics allows the material to give, enabling stresses to be distributed over an area, rather than be concentrated at a point. In most instances woven fabrics are employed although sewn fabrics are known to have a greater resistance to tearing. In sewn fabrics the fibres are loosely laid out longitudinally and transversely to the direction of fabrication, and are sewn together with thinner threads. The tearing resistance is improved, since the internal stress conditions within the fabric are considerably reduced. Two-directional fabrics can only develop their maximum strength in the longitudinal and transverse directions, and so tear resistance is comparatively low. The logical development was therefore to build-up fabrics with several plies, the warp and weft of each successive layer being angled to the next. This multi-ply construction produces highly compacted fabrics of extreme strength. However, such compaction requires the employment of additional adhesive agents when coating the fabric. With fabrics of an open structure, particularly sewn fabrics, mechanical adhesion with the coating is good, and no additional adhesive agents need be used in the coating.

Coated Fabrics : the Elastomer Coating

The vinyl products are by far the most popular elastomers used for coating fabrics, due in the main to both their cheapness and versatility. Its great advantage is that its composition can be altered to suit different requirements, although the best of all worlds cannot always be achieved, i.e. extreme cold flexibility can only be achieved at the expense of flame resistance and coating adhesion. However, it is the more expensive elastomers such as neoprene, butyl, and Du-Pont 'hypalon' and 'viton' compounds, which have superior weathering properties. Neoprene and butyl, both synthetic rubber compounds, have been in use for quite a long time, particularly for inflatable survival life-rafts, in which they have proved themselves as acid, oil and flame resistant, remaining flexible at low temperatures, durable and possessing extremely good weathering characteristics. Hypalon as well as possessing these characteristics can be compounded in a wide variety of colour fast hues. Its performance on the early DEW line radomes has been admirable and recent improvements in its make-up have increased the retention of its physical properties, as witnessed by its performance on the more recent and larger 'Telstar' radomes. In addition this hypalon can easily be recoated to extend the life of the fabric even further. Although the coating is primarily there to protect the fabric and maintain its strength, the membrane material acts as a homogenous unit in which the coating is reinforced by the fabric. Since fabric deterioration can be caused by internal as well as external influences, the fabric must be coated on both sides with, of course, a higher percentage of the coating being applied to the outside surface. This percentage is often as high as 75 per cent but a figure of 60 per cent is more usual. This percentage is dependent on the method of jointing employed, as well as the required service life and the influencing factors of the internal environment.

Undoubtedly, the most favoured membrane material is vinyl coated nylon or terylene. Extremely good adhesion can be developed between the coating and its base fabric, an essential requirement of all coated fabrics to ensure maximum protection of the base fabric. At the moment a service life of between 5 and 10 years can be expected for this material, and indeed, many structures have been in use for over 10 years.

Coated Fabrics : the Jointing Methods

Jointing methods for the membrane material must be selected so that the full strength of the basic material is developed. Experience has shown that a simple lap joint provides a joint of high efficiency, although double folded seams are favoured when joints are sewn. For maximum efficiency, the shear load must be distributed uniformly across the joint. This distribution is dependent on the material's plastic properties. Materials with a high modulus of elasticity stretch very little, and consequently the movement at the interface of panels is small. This results in relatively uniform shear stresses in the material, and provides an efficient joint. Unfortunately, coated fabrics have a low modulus of elasticity which results in significant movement at the interface of panels. Therefore the distribution of shear across the joint varies from a very high stress at the edge to a relatively low stress at the centre. This condition can result in edge peeling, with subsequent joint failure, known as the 'zipper' effect.

Three methods of jointing are practised universally; high frequency welding or heat sealing, cementing and sewing. The method employed will depend on the type of material used, and on financial considerations. With vinyl coated fabrics welded or sewn joints are favoured. Cementing the joints is far more expensive and offers no advantage over the other two methods. Welding is the most satisfactory method, but it cannot be used if more than 60 per cent of the coating is on one side of the material. If the fabric is of open structure, there is little obstruction to free plastic flow, and welding fusion will extend throughout the full thickness of the material. Such a joint allows the full development of the material's strength and is very neat in appearance. Sewn joints are much cheaper, but are less efficient in terms of material strength development and service life. The strength of a sewn seam is about 75 per cent that of a welded or cemented seam. Normally a triple sewn seam or a double folded seam is used. In the latter, four layers of material lie one above the other and are double stitched. Threads which are resistant to ultra-violet rays, such as dacron or terylene, must be used, and this stitching must be sealed with a vinyl coating. If the latter is not undertaken, moisture can penetrate into the fabric via the stitches. Cemented joints are prone to the 'zipper' effect, and thus may require additional strengthening. This can be achieved by double taping of the joint, or even sewing. Despite the unquestionable superiority of the welded joint, sewn joints are very popular with manufacturers. The probable reasons for this are, firstly, cheapness and, secondly, the fact that with a sewn seam more than 60 per cent of the coating can be employed on the outside surface, thus giving greater protection to the fabric from the weather.

Fig. 59

Fig. 60

Where superior coatings are employed, such as neoprene, hypalon or butyl rubber, welding is unsuitable, consequently cemented or sewn seams must be resorted to. Cementing is a difficult process, and special equipment and fixtures are required to ensure a uniformly high strength bond, free of air pockets, voids and wrinkles. Although this method is much more expensive, it is appropriate for these superior materials. When glass fabrics are used it is advisable not to join the material by sewing, since the resulting load concentration may weaken

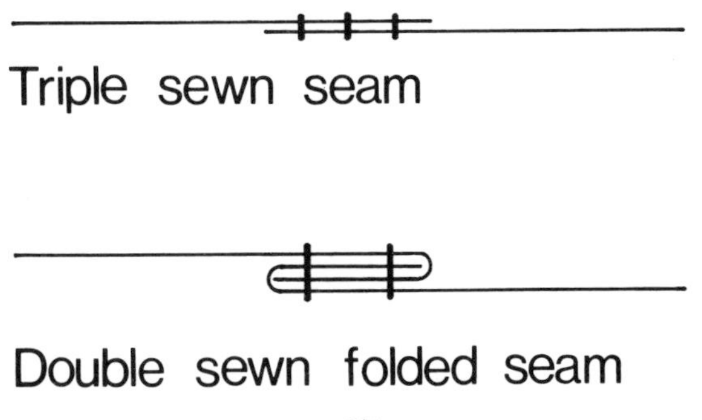

Fig. 59

Fig. 60

the fibres and cause failure. At present, much research is being undertaken throughout the world developing coated fabrics suitable for pneumatic structures. The following table gives an idea of the strength of some of the coated fabrics at present available, but development is such that these are being replaced daily with superior products.

Material	A	B	C	D	E	F
Hypalon coated dacron fabric (2 ply 45° bias)	U.S.A.	1·78	2380	910	910	180
Hypalon coated polyvinyl alcohol fabric	Japan	0·92	980	770	720	48
Silicone rubber coated dacron	U.S.A.	0·51	540	270	270	41
Vinyl coated polyester fabric	Sweden	0·8	950	490	410	60
Vinyl coated polyester fabric	U.K.	0·51	540	180	115	52
Vinyl coated nylon fabric	U.S.A.	0·66	610	360	305	50
Vinyl coated polyamide fabric						
1 fabric sewn	G.D.R.	0·76	760	390	375	—
2 fabric woven	G.D.R.	0·76	760	375	370	—
Vinyl coated polyamide fabric	Sweden	0·6	700	300	260	35

A Country of manufacture
B Thickness, mm
C Weight, g/m^2
D Longitudinal strength, kgf/50 mm width
E Transverse strength, kgf/50 mm width
F Tearing resistance, kilogrammes

In this comparison it should be remembered that testing methods vary throughout the world.

Woven Metallic Fabric Membranes

Prompted by the space programme, research is at present being carried out in America to establish the suitability of woven metallic fabrics for pneumatic structures in space and lunar applications. This research could quite probably produce a material also suitable for earth-bound pneumatic structures. At the moment stainless steel fibres, coated with a suitable material for airtightness, hold great promise. Such materials possess great strength and durability, but unfortunately cost rules them out as viable propositions for everyday use.

Metallic Foil Membranes

As with woven metallic fabrics, research into metallic foils was prompted by the NASA space programme. An aluminium plastic foil is already being utilised because of its high reflection of heat. It is lightweight, very strong, impermeable, and possesses good ageing properties. As yet its feasibility for large structures has to be demonstrated, but its properties, particularly its remarkable heat insulation properties, make it a very attractive proposition for future pneumatic structures.

Rigidising Plastic Foam

The need for maintaining an excess air pressure within pneumatic structures is considered by many as a major disadvantage of such structures. Consequently, rigidising foams, such as polyurethane, have been developed, which can be introduced into pneumatic structures to produce rigid forms that no longer require a continuous air supply for supporting purposes. However, once this rigidising element is introduced into the structure, the principal attribute of a pneumatic, its portability, is forsaken, and it can no longer claim to be a true pneumatic structure.

Membrane Supporting Elements : the Inflation Equipment

As yet, the only method for maintaining the stability of air supported structures, is by inducing air into the structure by means of fans. The inflation equipment is the 'heart' of an air supported structure and must be completely reliable at all times. It is as essential to the air supported structure as the engine is to the aeroplane. Due to pressure differentials, air will leak from the interior of the structure at every conceivable point. The three major points of leakage are, firstly through the door openings, secondly around the perimeter at the structure's junctions with the ground, and thirdly through openings provided for ventilation purposes. Air can even permeate through the membrane material itself and through the joints, especially if sewn methods of jointing are employed. However these latter air losses are usually negligible. Owing to the difficulty of estimating the magnitude of these losses, the calculation of the required fan capacity is largely an empirical process. The air leakage through door openings is un-

Fig. 61

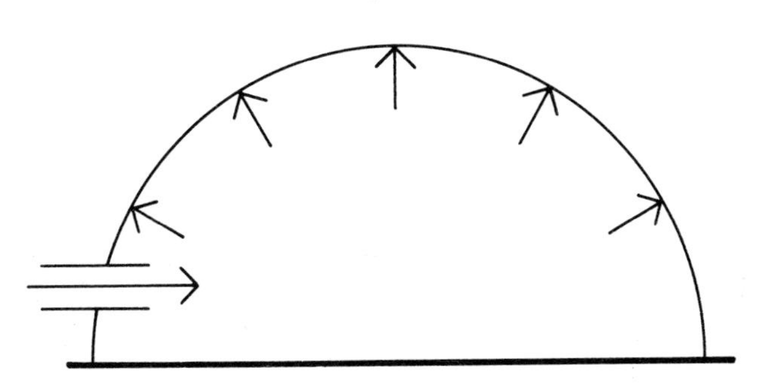

Fig. 61

doubtedly the most significant. As structures increase in size so the relative size of the door openings to the overall size decreases, and therefore the relative size of inflation equipment needed to maintain stability diminishes. With smaller structures the size of inflation equipment is generally such that the internal environment is sufficiently

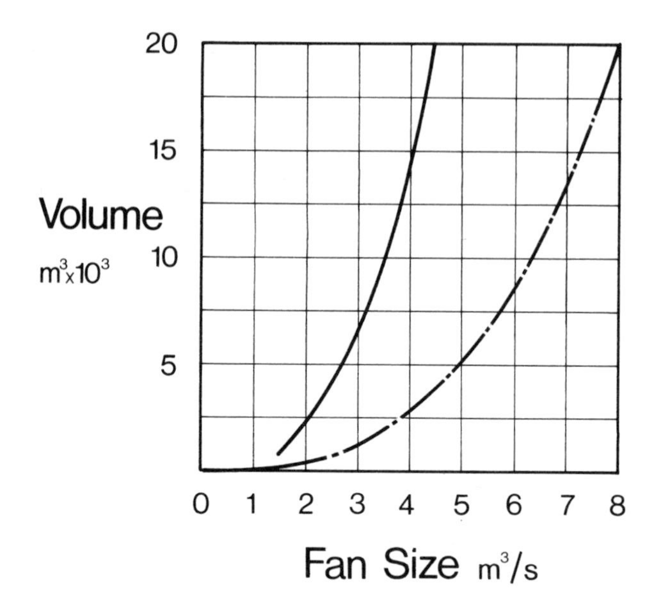

Fig. 62

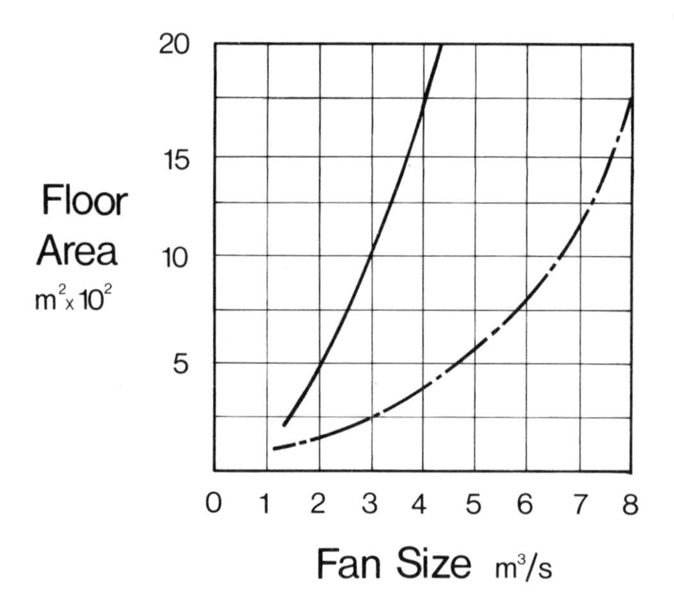

Barracudaverken ————
Kurashiki Rayon —·——·—

Fig. 63

ventilated. However, with larger structures ventilation rates are indeed minimal if the inflation equipment is only designed to achieve structural stability. The accompanying diagrams indicate the relationship between plan area and fan size, and between building volume and fan size, based on the equipment used in the standard production air structures manufactured by 'Barracudaverken' of Sweden and 'Kurashiki Rayon' of Japan. These comparisons illustrate how standards vary considerably throughout the world and also point out the fact that need for greater ventilation rates may determine the sizing of fans for larger structures. As a rough guide the fan capacity should be such that the air changes are between one and two per hour, but this is not really applicable to either very small or very large structures.

Fig. 62

Fig. 63

$$\text{Fan capacity in m}^3/\text{s} = \frac{K}{3600} \times (\text{Building volume, m}^3) \qquad (4.1)$$

where 'K' lies between 1 and 2. Since air leakage is not usually constant, depending on the amount of traffic in and out of the building, it is necessary that the pressure level in the building be automatically controlled either by variation of the air supply or by running the fan at maximum capacity at all times and allowing excess air to pass out of the building through a pressure relief flap. The former is much more sophisticated, but it is generally ruled out because of financial considerations.

Design Criteria for Inflation Equipment

The following basic principles are suggested for the provision of a reliable inflation system.

1 The capacity of the inflation system must be designed to cope with maximum critical conditions occurring during the initial inflation, peak loading periods, and emergency situations. For the majority of the time the inflation system will only be working at about a third of its maximum capacity.
2 Dual or multiple fan systems, with only one fan operating under normal conditions, provide additional reliability against mechanical failure. Back-draft flaps prevent air losses through inoperative fans.
3 A variable output fan permits manipulation of the internal pressure loading to correspond with changes in external loading due to climatic variations, as does a multiple fan system.
4 Snow must not be allowed to build up around the air intake, thus blocking off the air supply.
5 Due attention must be given to the design of the air intake to ensure that the air supply is not affected by wind turbulences around the structure. This can be achieved by positioning the air intake remote from the structure and away from any major aerodynamic influences, or by taking the air supply from a plenum chamber fed

Plate 35 An axial fan, cheap and simple, but rather noisy

from a number of ducts. The latter method ensures a minimum variation caused by changes in wind speed and direction.

6 An alternative power supply must be provided, and this must be automatically activated in the event of a primary power failure. In cases where only one fan is provided for normal inflation a completely independent air supply should be provided, such as a petrol or oil driven fan or alternatively an electrically driven fan powered by a stand-by generator. This auxiliary safety equipment must be frequently inspected to ensure that it will automatically operate when needed. As yet many structures do not have an alternative means of air supply. For uses such as warehousing it is often felt that a collapse due to air supply failure is of very little consequence provided the contents have no sharp protrusions that could rip the membrane. However, a high wind could cause extensive damage to an unstabilised membrane; therefore this practice is not recommended.

Unfortunately many users are still unwilling to pay for the additional

Plate 36 The centrifugal fan, more sophisticated than the axial fan

outlay involved in providing a reliable inflation system based on the above principles. It is hoped that legislation will remedy this fact and also help to eliminate some of the misconceptions that are held about the safety of air supported construction.

At present there are two main types of low pressure fans used for the inflation system, axial and centrifugal fans. The former, though cheap, compact, rugged and simple, are extremely noisy and therefore only really suitable for industrial applications. Attenuators used in conjunction with axial flow fans, have been reasonably successful in reducing the noise level. However, centrifugal fans are more sophisticated and quieter, making them more appropriate for pneumatic buildings.

Plate 35

Plate 36

Ideally the air should be supplied uniformly from around the periphery of the building, so that the occupants are not conscious of the large volumes of incoming air. This could be quite easily achieved with a peripheral ducting system having numerous outlets that discharge upwards. However, very few examples exist with this degree of sophistication.

Fig. 64

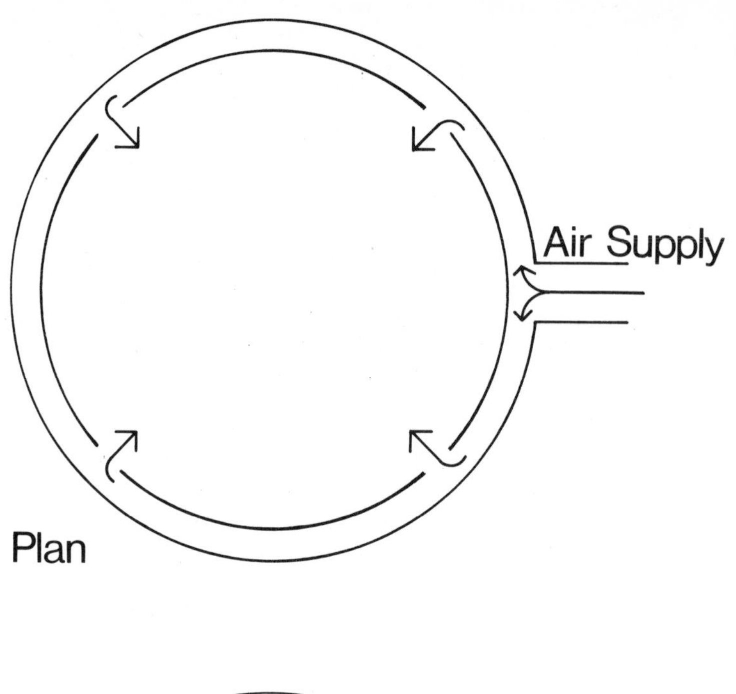

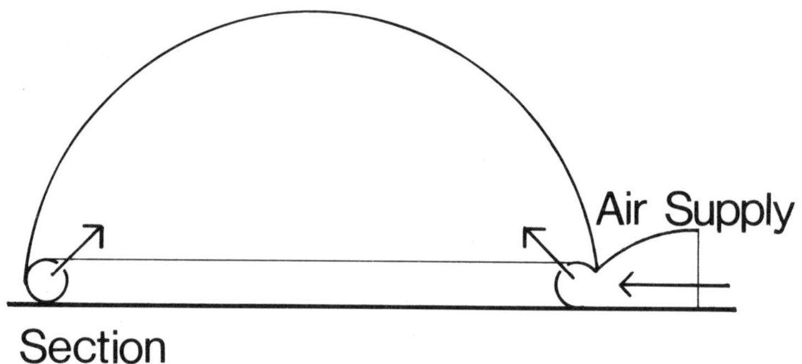

Fig. 64

Membrane Support Utilising the Wind

The wind is one of the major influences governing the required internal pressure, and it would therefore seem logical to attempt to harness the wind's energy to attain this internal pressure differential. As the wind velocity increases, a greater pressure differential is required for stabilization. This could be achieved by utilising the resulting increased wind pressure, making the structure self compensating for the varying wind conditions. Obviously such a system would need auxiliary inflation systems for windless conditions. The wind could be harnessed by means of collectors placed outside the structure's aerodynamic influence and directed to the interior. With shallow profile structures in which the wind causes only suction loads on the membrane, there would be no need for wind collectors as the suction forces would be sufficient to maintain the stability of the structure. In high winds, high stresses could occur in the membrane due to these suction forces, but these could be reduced by evacuating some of the air, so

Fig. 65

reducing the pressure differential. Since nature is so unpredictable, such methods would need careful investigation, before being employed.

Membrane Support by Upthrust Methods

In addition to these methods, the membrane can also be supported by the upthrust forces of hot air or lighter than air gases. With some of the larger air supported structures erected in colder climates, it is known that the structure takes a considerable time to deflate after terminating the air supply, because of the upthrust forces induced by the hotter air. Air, when heated, rises due to its decreased density, a phenomenon exploited by the early balloonists. By maintaining a temperature differential between the outer and inner environment, the contained warm air would exert an upthrust on the structural membrane. Whereas with a continuous air supply, the internal pressure differential acts uniformly in all directions at all points on the membrane's surface, the stabilising forces in the upthrust method act only vertically. Assuming a hemispherical dome the internal normal

Fig. 66

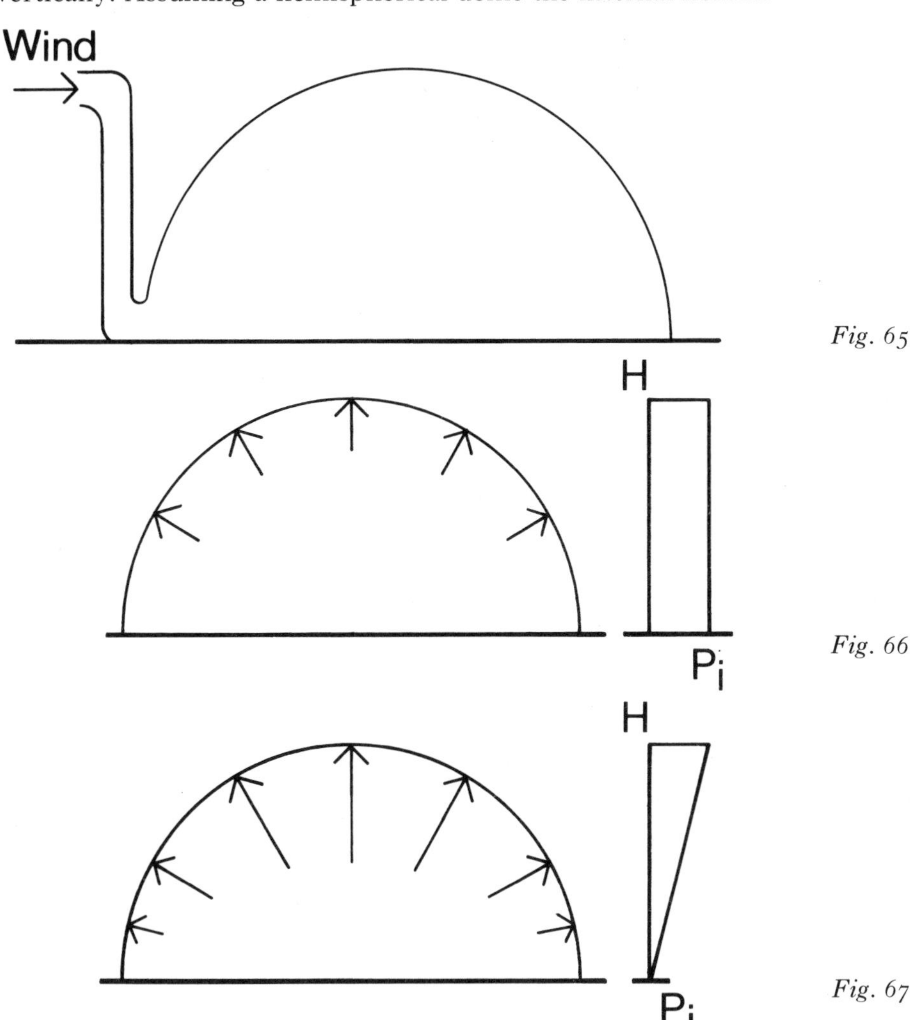

Fig. 65

Fig. 66

Fig. 67

Plate 37 Proposals for the enclosure of Wembley Stadium football pitch with a gas-filled balloon structure by architect Arthur Quarmby and plastics consultant David Powell

Fig. 67

pressure exerted on the membrane at the apex is a maximum, whilst at ground level it is zero. This has the great advantage that where air leakage is most likely to occur, through door openings and at anchorage points there is hardly any pressure differential to cause leakage. Consequently very little air replenishment is necessary. Unfortunately

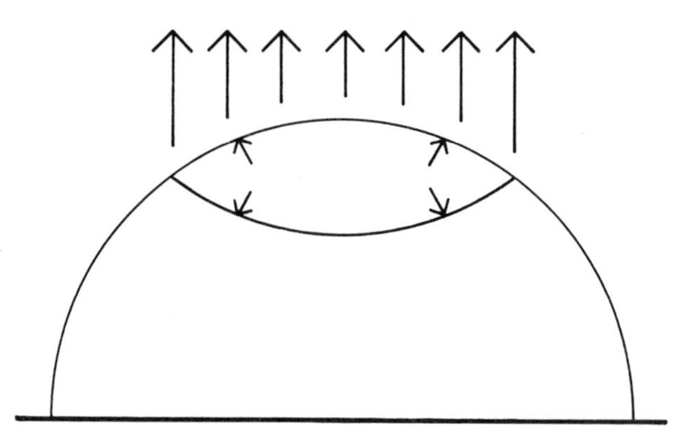

Fig. 68

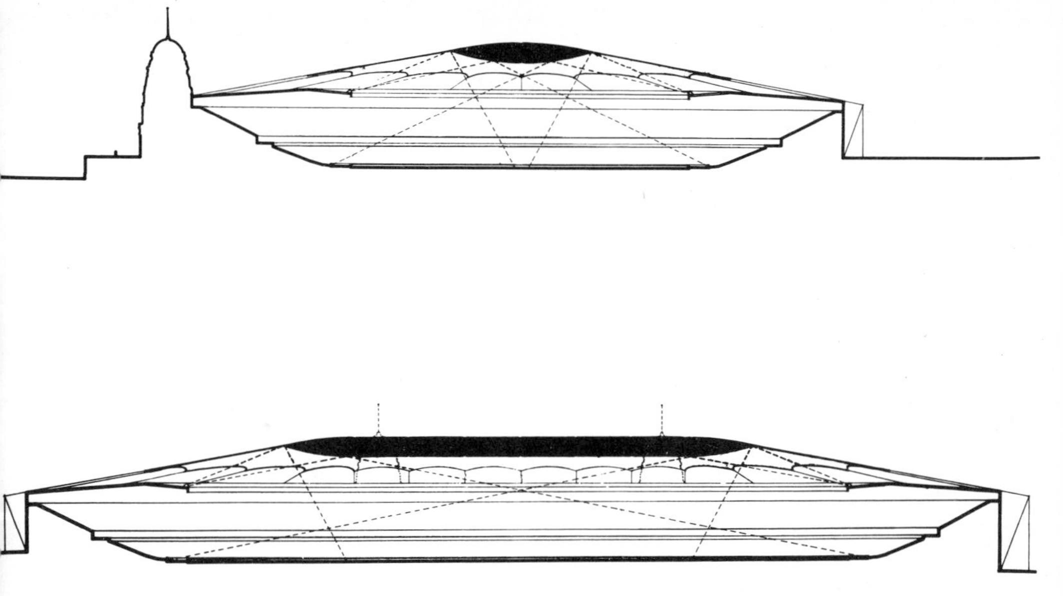

Plate 38 Sections through Wembley Stadium

this method only gives sufficient support on its own if the spans and temperature differences involved are both very large.

The alternative upthrust method utilises the lighter than air gases such as helium; the explosive nature of hydrogen rules it out as a feasible proposition, as the numerous airships destroyed by fire emphasise. These gases could be contained within the roof or apex of the structure, forming a floating 'pillow' structure. Arthur Quarmby, architect, and David Powell, a plastics consultant, have proposed this use of gas for covering in England's international football ground, Wembley Stadium; a membrane fixed to the perimeter of the stadium is supported in an elevated position by a helium filled balloon. Such ideas, although apparently practical, require much research work before being introduced. However, this sort of structure can hardly be called air supported.

Fig. 68

Plate 37

Plate 38

Anchorage Design

Pneumatic structures are unique in the manner by which they defy gravity. Unlike most conventional structures whose weight must be distributed into the ground, pneumatic structures impose an uplift load. When this uplift is not adequately opposed by a suitable anchorage, the structure, or some part of it lifts away from the ground, allowing air to escape, thus leading to possible structural failure. Whatever method is used for anchoring the structure, the anchorage forces must be distributed uniformly around the perimeter of the structure, to avoid any stress concentrations in the membrane, unless the membrane is reinforced by cable and nets, in which case point anchorages will be necessary. There are two methods of opposing these lift off forces, by ballast, and by positive anchorage to the ground.

Fig. 69

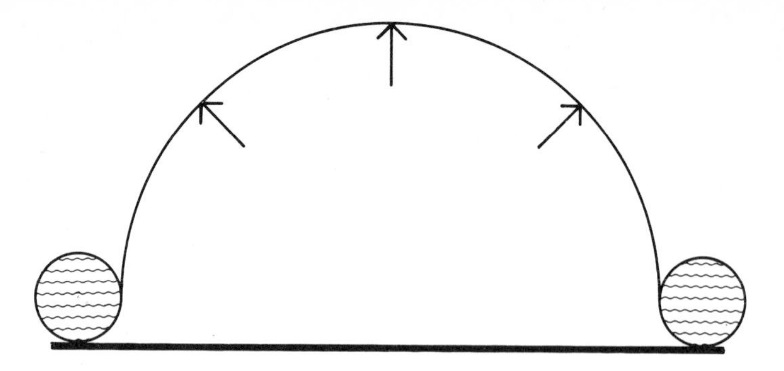

Water Ballast Anchorage Systems

Ballast anchoring is mainly used for structures that are frequently moved from one site to another. Since site conditions can vary considerably ballast methods may be appropriate. On first reflection the use of water ballast would seem ideal, since water costs little and is readily available. However, a continuous tube of water is vulnerable to vandalism as well as accidental damage, and it only needs a single hole to render the structure unsafe. If loading conditions produce high local lift off forces, the water will tend to flow away from this point, leaving it locally with no anchorage provision. For this reason the tube must be compartmented so that local failure does not affect the stability of the whole structure. For safe anchorage, very large quantities of water are required and this generally proves to be too cumbersome.

Fig. 69

Earth Ballast Anchorage Systems

Another method is to fill the ballast tube with solid matter, such as earth, sand or gravel. The ballast tube is attached to the perimeter and is split along the outside to facilitate filling and emptying. When the tube has been filled it is laced or zipped up so that it is firmly attached to the structure. Another variation of this method is to dig a perimeter trench, into which the surrounding skirt of the structure is placed; subsequent backfilling of the earth over the skirt provides anchorage. Denser ballast, such as concrete slabs and stones, are also suitable. Here again, the ballast is placed on the surrounding skirt. Although ballast anchorages were very popular when air structures were first introduced in the 1950's, Walter Bird was very wary of such methods, since the high forces on the anchorage, encountered under conditions of maximum loading, made them impractical for most applications. Several early failures of ballast anchorages occurred thus reinforcing Bird's scepticism. Ballast methods should only be used for structures of a very temporary nature where ground anchorage is an impractical proposition. When these methods have to be used, extreme caution must be exercised.

Fig. 70

Fig. 71

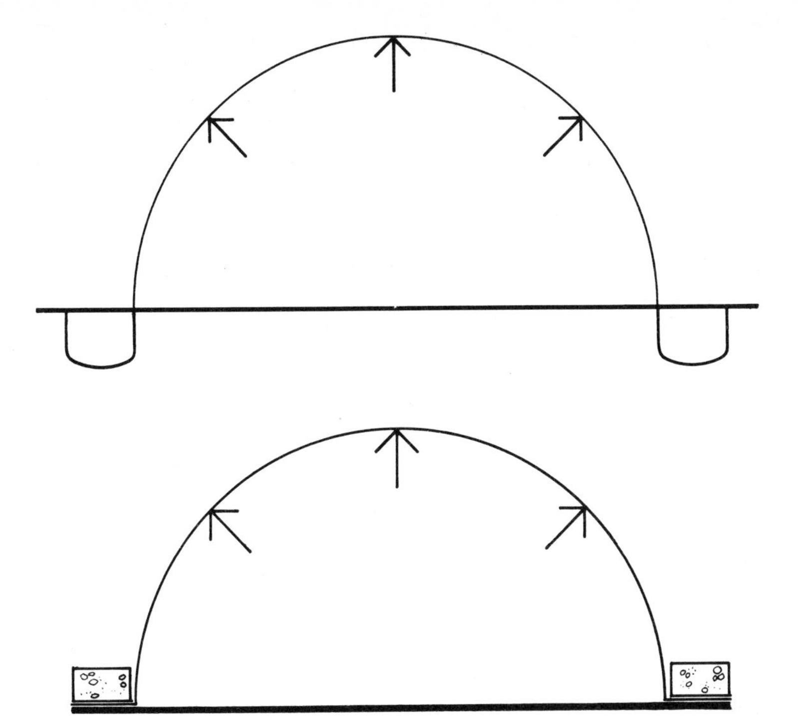

Fig. 70

Fig. 71

Ground Anchorage Systems

Ground anchorage systems are now used almost universally, the membrane being positively attached to the ground at frequent intervals. The anchorage forces must be uniformly distributed into the membrane around the perimeter of the structure, and this can be accomplished in many ways.

With clamped anchorages, a rope welt is usually sewn round the bottom edge of the membrane material, and steel channels, angle irons, pipes or even timber grounds, are bolted through the fabric just behind

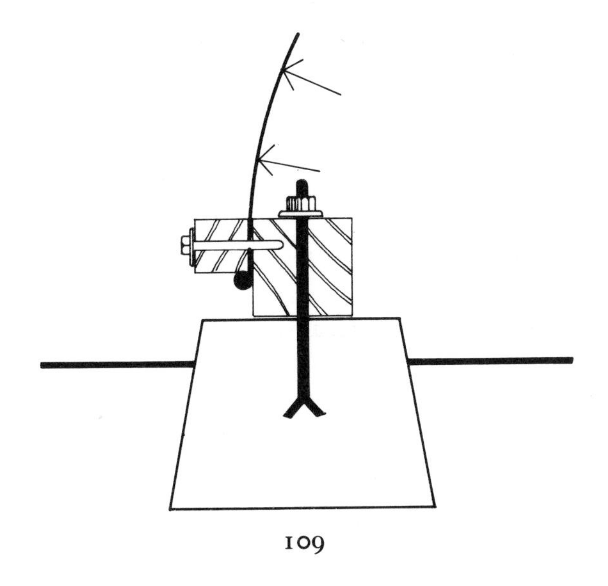

Fig. 72

Plate 39 Pipe-in-hem anchorage

Fig. 72 the welt to flush anchors. These steel or timber sections form a continuous band around the perimeter, and in turn are attached at approximately 1 m centres to the flush anchors, which are generally set in concrete.

The 'pipe-in-hem' anchorage consists of an open hem, sewn round the bottom edge of the membrane, with semi-circular cut outs at approximately 1 m intervals. Pipe sections are inserted in this hem and are attached to anchors at the cut out points. Quite often steel cables or ropes are employed in the hem, instead of the rigid pipe sections, but these are not as satisfactory, since the uneven distribution of the anchorage forces into the membrane causes local stress concentrations. Despite this, cables are frequently employed because of their cheapness.

Fig. 73

Plate 39

In catenary cable anchorages, a cable or rope in catenary form, is located inside a fabric sleeve, which is sewn to the base of the membrane material. This catenary is also attached directly to anchors, at frequent intervals. Such a system is very flexible and is particularly useful where a structure is to be used on two or more sites with permanently set anchors, or where a structure is to be erected seasonally in one location. Here again, local stress concentrations can occur at the end of each fabric sleeve where the catenary arcs clip onto the anchors. The

Fig. 74

exact positioning and tightening of the cable is extremely critical. Deep catenary arcs, semi-circular or greater, present few problems, but shallow arcs often cause serious difficulties.

Although clamped anchorages are most reliable, they do not possess the versatility of either the 'pipe-in-hem' or catenary anchorage systems, neither of which need a concrete foundation. In all methods, the main requirement is to keep the air leakage to a minimum, and to achieve this, skirts are attached to the inside, at the base of the structure. The action of the pressure differential on this skirt is sufficient to form an efficient air seal against the gound.

In all the above methods the membrane material is attached to ground anchors, which can be loosely classified into two groups, surface and underground anchors.

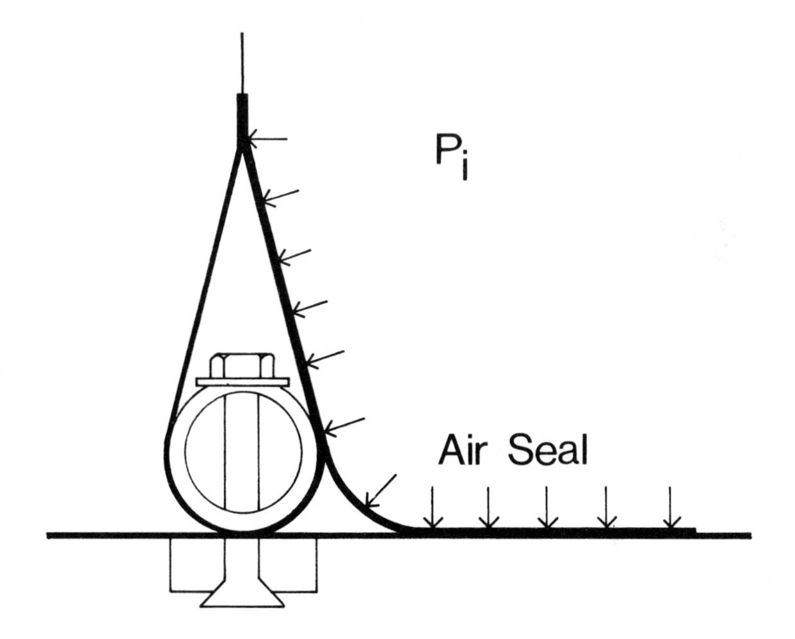

Fig. 73

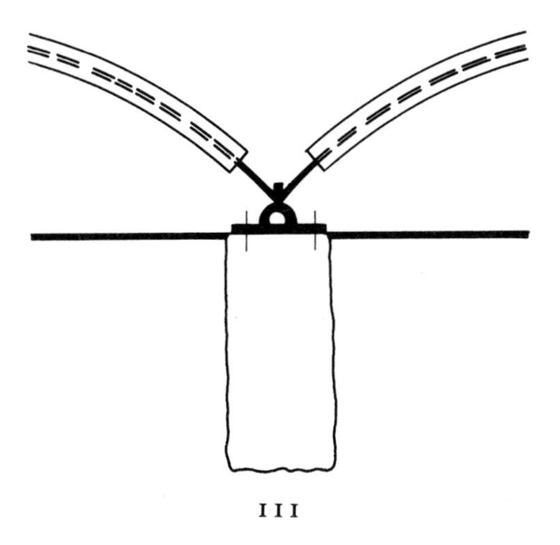

Fig. 74

Surface Ground Anchors

The most common method of anchoring the air supported structure to the ground is to attach it to a concrete foundation with flush surface anchors, such as ragbolts. When the structure is installed on existing concrete, steel anchors are inserted by electric impact hammers or by drilling and grouting in the anchors. Self-drilling expansion anchors can also be used. With new concrete foundations such methods can also be used, once the concrete has set. However, a much more satisfactory anchorage is accomplished by attaching the anchors to the reinforcement rods before the concrete is poured.

Underground Anchors

For underground anchorages spiral screw, helical plates, spreading and driven arrow-head anchors would appear to be suitable. However, when a great number of anchors have to be positioned accurately, underground obstructions such as stones often considerably hamper the installation. One of the most efficient and reliable of these anchors is the steel expansion type. This is installed in augered ground holes, about 0·25 m in diameter and 2 m deep. Underground obstacles pose few problems for augering equipment. The anchors are placed in the holes and expanded into the undisturbed earth. The holes are then back-filled with consolidated earth, leaving a neat threaded steel stud or flush internally threaded socket at the surface, for connection to the structure's anchorage points. Vertical underground concrete columns are also very satisfactory, providing rather a more permanent type of anchorage. Concrete is poured into augered holes of similar size, and a flush threaded insert is cast into the top of the column for attachment

Fig. 75

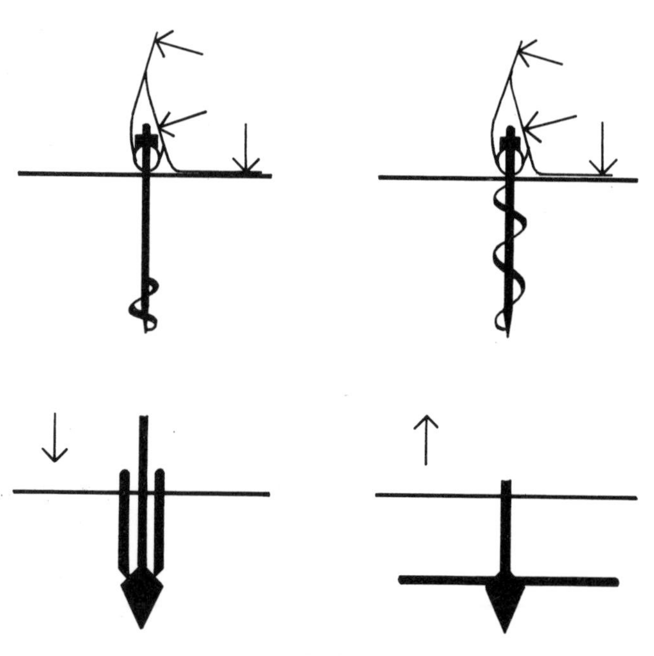

Fig. 75

to the structure. Since the anchorage is derived from the weight of the column, variations in soil conditions do not affect its anchorage magnitude very much; 'belling' the bottom of the hole greatly increases the holding power in certain subsoil conditions. It appears that the anchorage systems are undoubtedly the most developed element in air supported construction, but it must be remembered that the problems they present are by no means new to the building industry.[1]

Openings for Access

Since a small pressure differential is needed to support the membrane material, air losses readily occur through door openings. It is this aspect of the air supported structure which presents a major design problem, but with careful thought, suitable provision can be made to reduce these air losses to a minimum. Although air locks are the most satisfactory solution, they are not essential for small personnel entrances, since the inflation system can easily accommodate the added load placed upon it. The simplest door, a throw-back from the tent, is the squeeze through opening, but this is very crude and unsatisfactory, since it continuously flaps as air seeps out through it. This makes it suitable only for domestic structures, such as covers for garden swimming pools.

Fig. 76

Because of pressure differentials the load on a conventional personnel door would be between 50 and 100 kg. If opening inwards, this would require a strong shoulder, and is therefore somewhat impractical. If opening outwards it would pop open as soon as it was released, endangering the user and also the structure, because of the difficulty in closing it. If counter-balanced by springs or weight to facilitate operation against this pressure, it would be difficult to operate when the pressure dropped in any emergency situations. The unsuitability of conventional doors for air supported structures is therefore apparent. However, one adaptation of a conventional door has met with great success. The door is counter-balanced by a vane which reacts to the same internal pressure as the door itself, thus counter-balancing the load on the door. This door needs only a light

Fig. 77

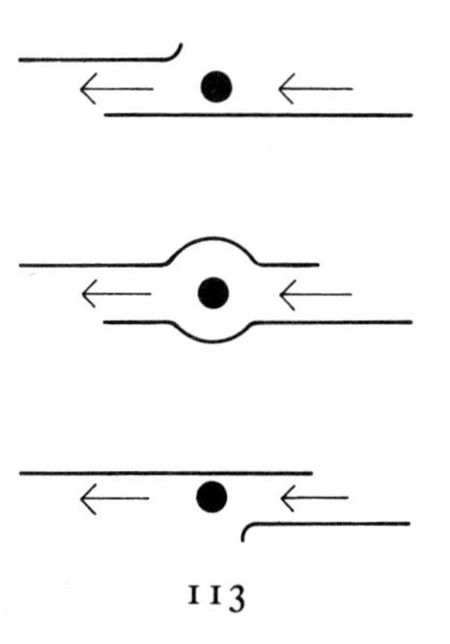

Fig. 76

spring to make it easy and convenient to operate, extremely safe, and very suitable for emergency exits.

When heavy personnel traffic is anticipated, the above solution is not acceptable, since the door would be more or less continually open, thus permitting huge air losses. For these heavily used applications the revolving door is ideal, not only because it is inherently an air lock, but also because people are familiar with its use. In many respects it is more suitable than an air lock, which relies on the common sense and discipline of people to close one door before they open the other. This becomes awkward and is frequently neglected under heavy use. Unfortunately revolving doors are not usually considered by building authorities as suitable for escape in the event of fire. This is because it

Fig. 78
Plate 40
Plate 41

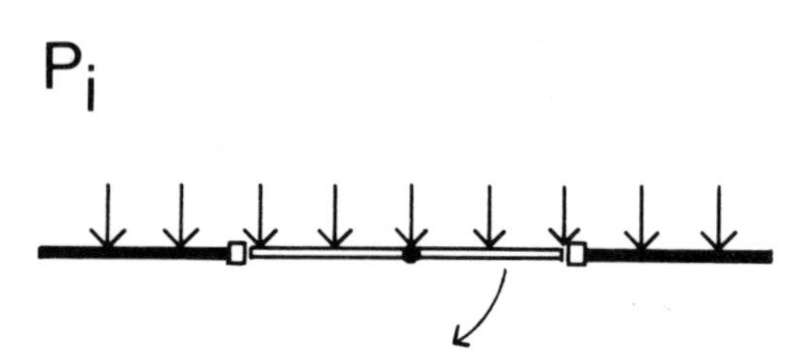

Fig. 77

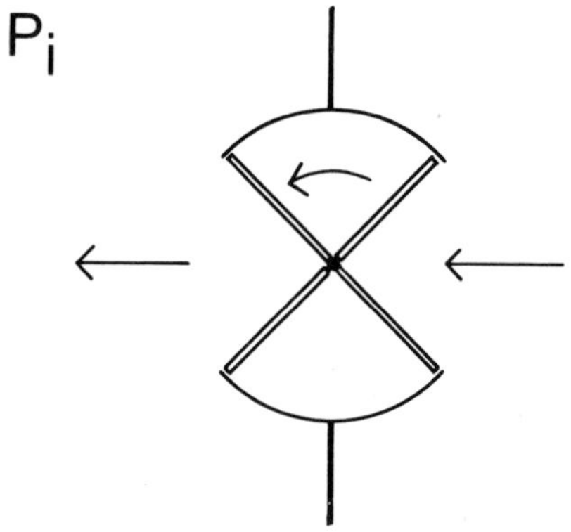

Fig. 78

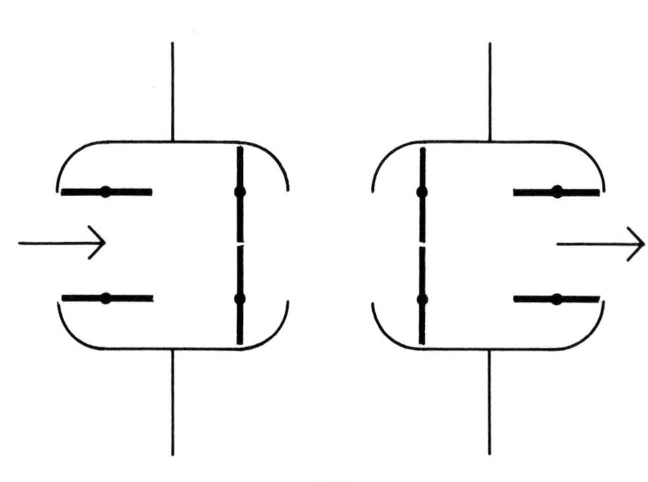

Fig. 79

Plate 40 Revolving doors for access

Plate 41 Typical personnel entrances on a Birdair structure

takes a considerable time for a large number of people to pass through a revolving door, and consequently an alternative means of access must in most cases be provided. Perhaps this can be in the form of a pair of vertical zipper openings at either side of the revolving door. For increased capacity the vane counter-balanced door used in an air lock system, as illustrated, is more appropriate and works quite well, although the air losses through this system can be quite large. *Fig. 79*

The larger openings required for vehicular access can only be satisfactorily achieved at the moment by means of an air lock or an air curtain. The former consists of a tunnel, big enough to contain the *Fig. 80*

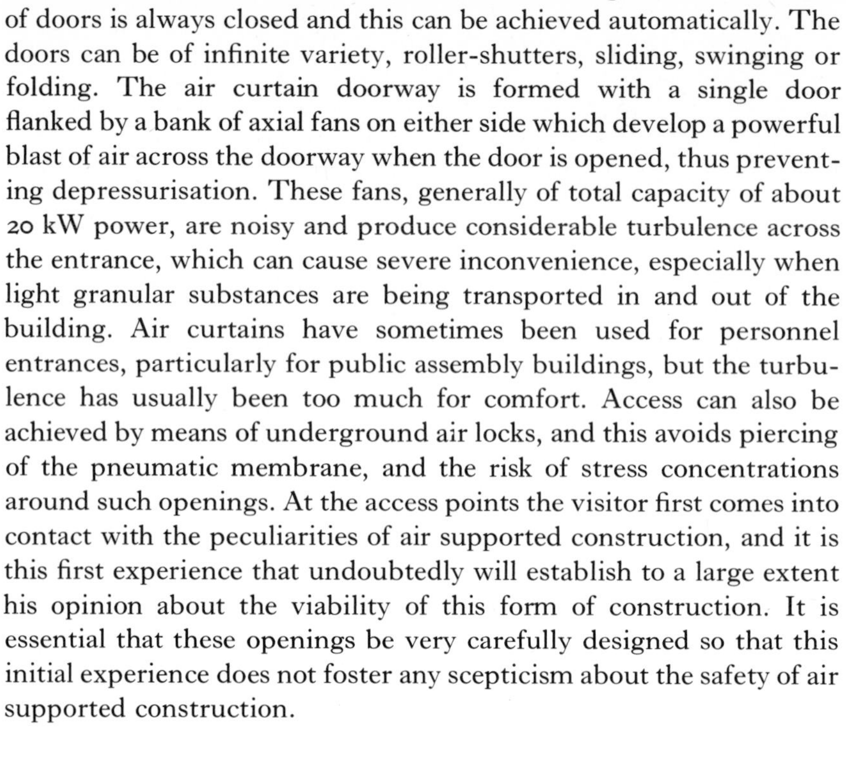

vehicle, with doors at either end; this must be designed so that one set of doors is always closed and this can be achieved automatically. The doors can be of infinite variety, roller-shutters, sliding, swinging or folding. The air curtain doorway is formed with a single door flanked by a bank of axial fans on either side which develop a powerful blast of air across the doorway when the door is opened, thus preventing depressurisation. These fans, generally of total capacity of about 20 kW power, are noisy and produce considerable turbulence across the entrance, which can cause severe inconvenience, especially when light granular substances are being transported in and out of the building. Air curtains have sometimes been used for personnel entrances, particularly for public assembly buildings, but the turbulence has usually been too much for comfort. Access can also be achieved by means of underground air locks, and this avoids piercing of the pneumatic membrane, and the risk of stress concentrations around such openings. At the access points the visitor first comes into contact with the peculiarities of air supported construction, and it is this first experience that undoubtedly will establish to a large extent his opinion about the viability of this form of construction. It is essential that these openings be very carefully designed so that this initial experience does not foster any scepticism about the safety of air supported construction.

Plate 42

Plate 43

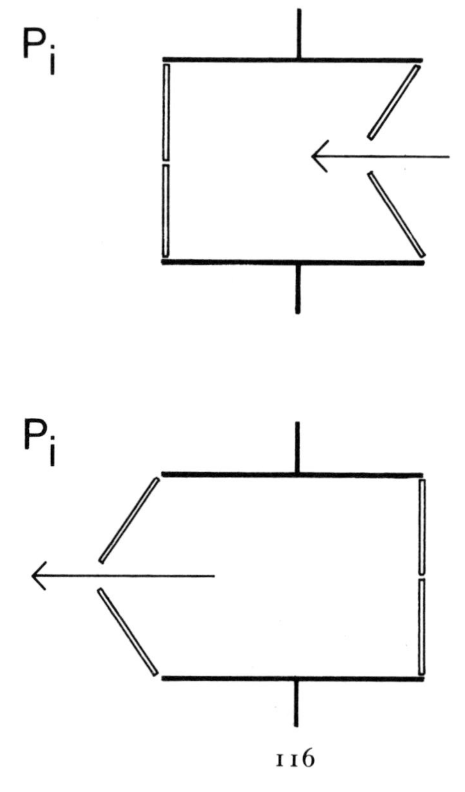

Fig. 80

Plate 42 Typical vehicle airlock on a Gourock air supported structure

Plate 43 An air curtain doorway on a Barracuda air supported structure

Special Design Problems

With air supported structures special attention must be given to the design of certain features where load concentrations are likely to predominate. The problem of the distribution of anchorage loads into the membrane has already been discussed. Other features requiring special attention are such things as cable attachments, where a point load has to be uniformly distributed into the membrane, membrane reinforcement around openings for access and ventilation, and the attachment of the membrane to rigid elements. In addition to this the form of the structure must be carefully chosen to avoid development of stress concentrations. Since pneumatic structures are dynamic structures, in which loads are balanced and redistributed by membrane distortion, rigid elements will restrict their movement and induce stress concentrations large enough to cause membrane failure. This can be avoided by using catenary cables to isolate the rigid elements from the membrane, similar to the manner in which vibrating machinery is isolated by flexible mountings. The stress is taken uniformly by the catenary cable, and the unstressed material between the cable and rigid element allows movement of the structure to take place without stress concentrations. Membrane failures have been most frequent around entrances, due to the negligence of designers in affording special attention to these design features. If stress concentrations are unavoidable then the membrane must be suitably reinforced at these points. Such reinforcement strengthens the overall structure against deformation.

Plate 44

With large structures, the membrane becomes very cumbersome to handle, unless it is sectionalised. Where this is necessary the stresses between adjacent sections must be uniformly transferred. This can be

Plate 45

Plate 44 Catenary cables isolate the membrane from rigid entrance structures

Plate 45 A Barracuda air supported structure fabricated in three sections

achieved by zippers, but unfortunately these have proved unreliable in service. Cord lacing of grommeted edges results in severe stress concentrations, and consequently this method can only be used for very lightly loaded structures. On the other hand, braid lacing is much more satisfactory, producing considerably less distortion, particularly if it is doubly laced.

Plate 46

Cable, Cable Net, Indent and Internal Membrane Wall Construction

So far very few structures have employed cable nets to relieve the membrane stresses, although single cables are quite often used to reinforce the membrane material. The limited experience with cable nets has shown that the abrasive interaction between cable and membrane can present severe chafing problems. It is therefore essential

Plate 46 Detail of the joint between sections of a Barracuda air supported structure

that designers look back to the days of the great airships and balloons for guidance on cable net design.

Although shallow profile structures can be achieved over large spans, with or without cables and nets, the resulting large membrane curvature gives rise to very high stresses. This problem can be overcome by anchoring the membrane to the ground, in the centre as well as at the span's perimeter: the resulting smaller curvatures give rise to reduced membrane stresses. Such indents can be arranged both regularly and at random, to derive various spatial forms. These indents form traps for rainwater, unless the overall slope of the roof is steeper than the line through the low point of the indent. It may thus be necessary to use the ground tie as an internal drainpipe. Alternatively, the water could be stored within the indent, and constitute a ballast anchorage, although it is unlikely that this would be visually acceptable. Indents produce internal structural obstructions, but these tensile members are nowhere near the magnitude of the compressive column members, in more conventional constructions. As well as point anchorages of indents, membrane walls similar to those occurring in soap bubble conglomerations can be employed. Like

Fig. 81

Fig. 82

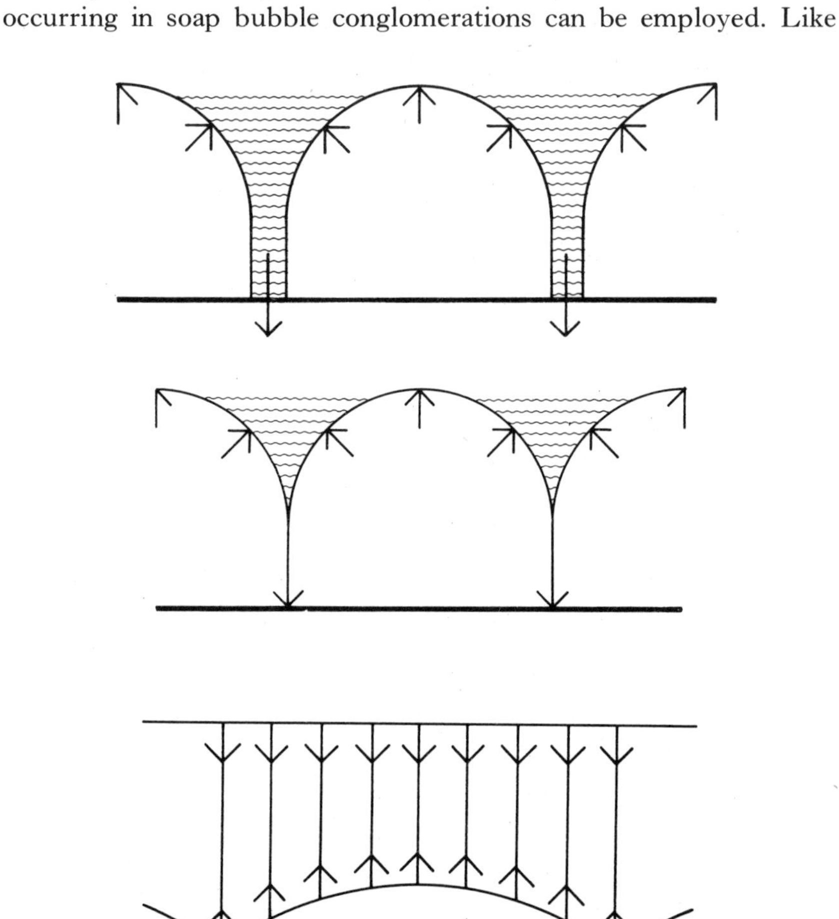

Fig. 81

Fig. 82

Fig. 83

Plate 47 A Barracuda air supported structure arrives on site

cables, vertical internal membrane walls, spanning between the roof and the ground, produce deep grooves in the roof membrane, thus providing drainage channels. Internal walls not only subdivide the pneumatic interior, but also reduce the span and stiffen the entire structural membrane. If the internal space must be unobstructed by walls, these walls can be reduced to cable tie arrangements, taking on the form of an inverted suspension structure. This type of construction does not impose the problem of internal drainage, common to indent construction. *Fig. 83*

Erection and Dismantling Procedure

Many claims are advanced about the ease of erection and dismantling of pneumatic structures, but how true are these? An examination of a typical air supported structure, of floor area 1000 m², suitable for storage or indoor sports, will help clarify this matter. *Plate 47*

First of all the foundations, whether in the form of a concrete slab or a concrete ring beam, must be prepared. The anchorage connections are positioned either before or during the placing of the concrete, the latter being most satisfactory. Entrances are then erected, and the structure's membrane material is attached to the fixing points. The inflation equipment is installed and all doors are closed. On activation *Plate 48* *Plate 49* of the fans the structure slowly starts to inflate and with this size of structure complete inflation generally takes between 45 minutes and an hour, although with additional fans it can be quite easily achieved in less than 30 minutes. It is difficult to predict this inflation time but the following expression gives an approximate idea.

Plate 50

$$\text{Inflation time} = \frac{K}{60} \times \frac{\text{(Volume of building)}}{\text{(Maximum fan capacity, m}^3\text{/sec.)}} \text{ minutes}$$

where 'K' is between 1·25 and 1·5 (4.2)

Plate 48 Drilling the concrete foundation to receive an expansion anchoring bolt

Plate 49 A Barracuda air supported structure ready for inflation

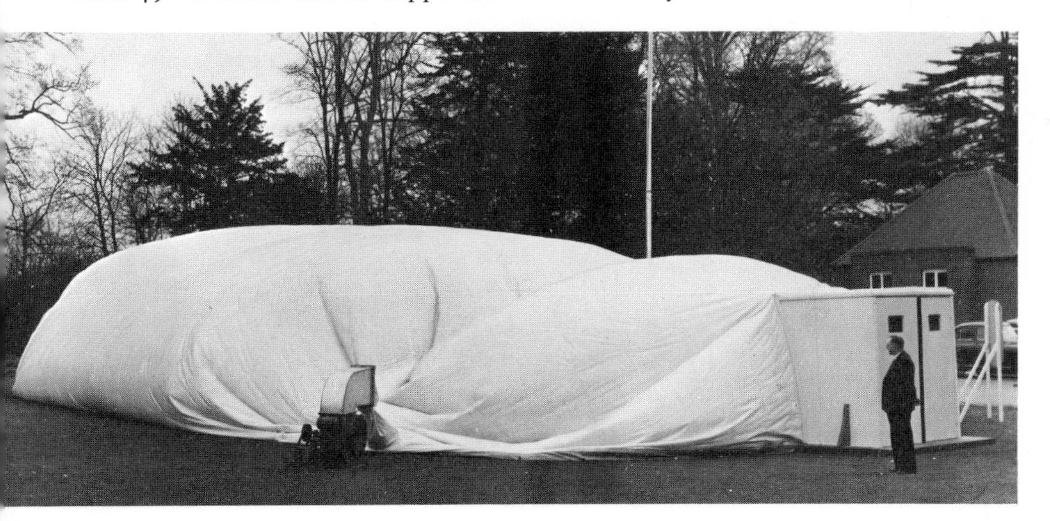

Plate 50 A Gourock air supported structure being inflated

It is essential that this inflation and also deflation is undertaken during periods of low wind speeds, since extensive damage could occur if the membrane was allowed to lash about uncontrollably. For this reason it is advisable to accomplish inflation as quickly as possible, and quite often additional fans are used just to speed up this operation. After inflation any catenary cables must be adjusted to equalise stress conditions, and anchorage and access points must be checked to ensure the safety of the structure. Complete erection on an already constructed concrete foundation can be quite easily achieved in 4 to 5 days by four unskilled labourers under the supervision of a skilled engineer. Once people have obtained the know-how they can be left to erect and dismantle their structure at will.

Dismantling is a much quicker and simpler procedure. The fans are switched off and all doors are opened, so that deflation is as fast as possible. Depending on the number of openings in the structure complete deflation generally takes between 5 and 20 minutes. Dismantling and packaging of the structure can be quite easily accomplished in less than a day by the same team of five men. Here again it is advisable to dismantle during periods of low wind speeds. With these simple structures neither erection nor dismantling presents any problems, but this is not usually so with more complex structures, each with erection and dismantling complications unique to themselves, and some of these are discussed in the following chapter.

Environmental Characteristics and their Control

In the introduction air supported architecture was described as a complete reversal of traditional architectural thinking, in that the application of environmental energy structurally stabilises the air supported building. However the true significance of this statement is as yet only realised in a very few pneumatic buildings. At present environmental control in pneumatic structuring on the whole lacks any degree of sophistication, although some developments hold much promise for the future.

The Thermal Environment

With single membrane structures, the membrane is extremely thin and provides very little insulation. Because of this, close temperature control is very difficult. Besides the large heat losses that occur these structures are vulnerable to the 'greenhouse effect', and these solar heat gains are known to produce temperature rises as high as 10 deg. C. To avoid large variations in the internal environment either additional insulation is needed or sophisticated air conditioning equipment must be installed. For most applications in which pneumatics are at present used, the latter method is economically prohibitive, initial costs of equipment being as much or more than the cost of the structure itself, and the operational costs are correspondingly high. In addition to this, such sophistication is in many cases not appropriate. Because of the low mass of pneumatic structures, virtually only the air within the structure has to be heated. This means the environment can be heated very quickly with short-time-lag heating systems appropriate for intermittent use, the building only being heated whilst it is occupied. Hot air and radiant systems are the most suitable, the former being incorporated into the pressurising equipment, i.e. the application of environmental power. Generally these systems are very crude indeed; in some cases the incoming air is heated with propane gas burners, but this is most ill-advisable since condensation on the inside of the membrane is most apparent. Indirect heating systems, or electrical heating equipment are the most suitable, as in these the air supply is passed through a heat exchanger.

Only in a few cases is cooling equipment used to extract the heat build up due to the 'greenhouse effect'. Consequently if these uncomfortable environments are to be avoided the solar radiation must be repelled. For this reason white membranes are very popular; recently developed aluminiumised materials have also been very effective when used. A cheap and surprisingly effective cooling system can be achieved by sprinkling the membrane with water. Perforated hoses are evenly distributed over the membrane's surface, and when attached to a mains supply, water is sprinkled over the membrane. The resultant surface evaporation of the water causes cooling of the interior of the building. The insulation properties of the structure can be greatly improved by using double membranes, between which are contained air or other insulation materials. Alternatively insulation materials, such as expanded polystyrene, can be attached to the interior of the membrane.

Air losses from small air supported structures through openings and joints generally provide sufficient ventilation of the internal environment. If this is not enough, as is often the case with larger structures, further air leakage can be induced by the provision of vents, thus increasing the ventilation rates. These vents are best positioned at the crown of the building, where stagnant hot air tends to accumulate; this of course entails increased running costs. Alternatively, an impression of higher ventilation rates can be accomplished by careful positioning of inflation equipment so that air movements, barely noticeable to the occupants, are induced within the building at the level of the occupants.

Laing's Pneumatically Controlled Radiation Walls

The preceding paragraphs illustrate in how few cases air supported construction has been realised as the application of environmental energy to achieve structural stability; there is no doubt that in the future more sophisticated pneumatic structures will achieve this realisation. However, this application of environmental power in pneumatics involves the use of terrestrial energy sources, and it is this fact that encouraged Nikolaus Laing to pursue the possibilities of controlling solar and terrestrial radiation pneumatically to achieve man's environmental needs. He described the results of his investigations in a breathtaking paper at the I.A.S.S. Pneumatic Colloquium in Stuttgart in May 1967.[2]

> 'The novel procedure is based on the principles of insolation and terrestrial radiation by controlling the emission and reflection of outside walls of the proposed new type of buildings. It will be

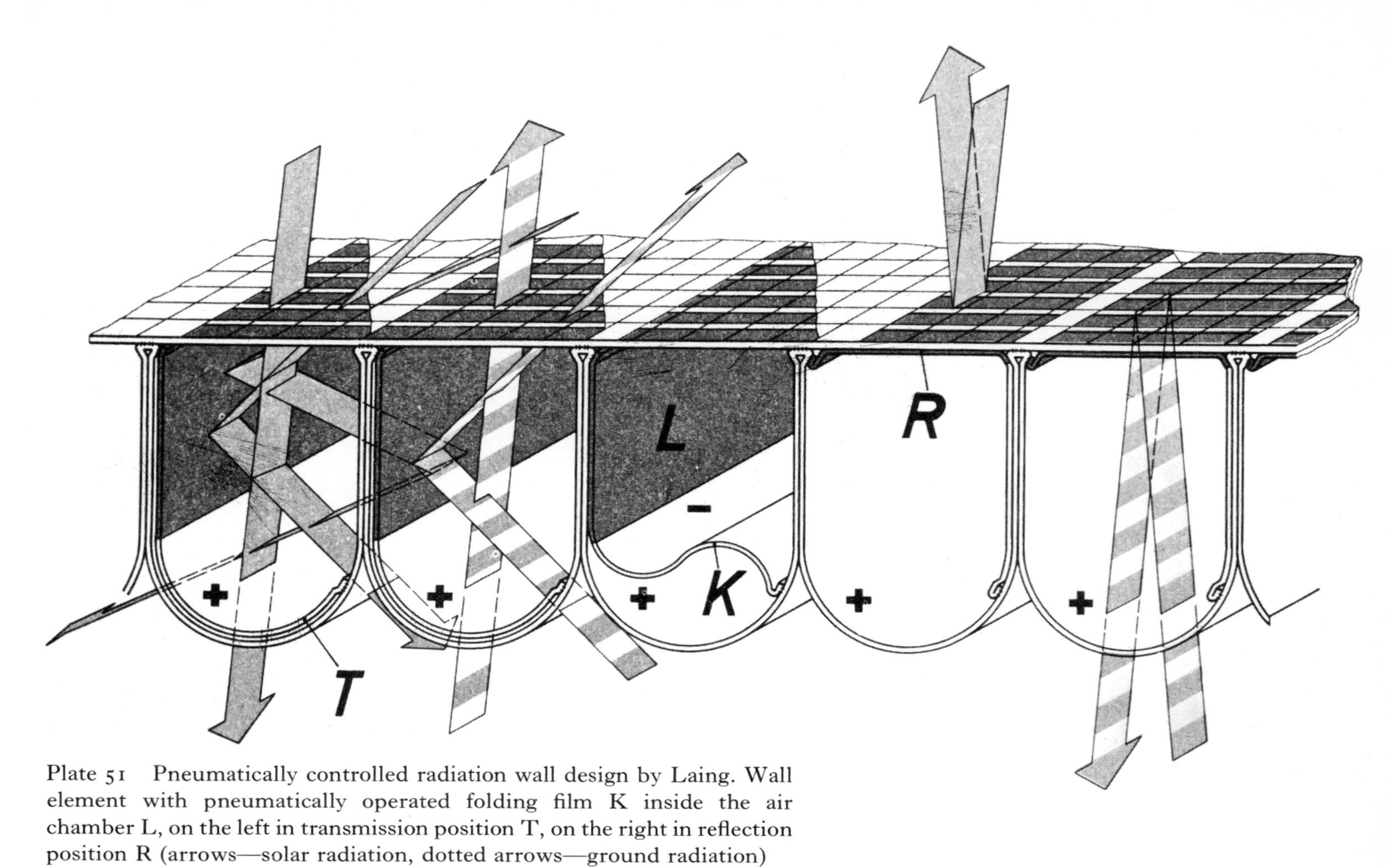

Plate 51 Pneumatically controlled radiation wall design by Laing. Wall element with pneumatically operated folding film K inside the air chamber L, on the left in transmission position T, on the right in reflection position R (arrows—solar radiation, dotted arrows—ground radiation)

shown that temperatures near zero °C in the Sahara and sub-tropical conditions in Newfoundland can be achieved by means of novel wall designs, combined with radiation and transmission control. Except for negligible amounts of energy for controlling purposes, the new system works without using any terrestrial energy sources.'[3]

Such is the sophistication of Laing's proposals, that they deserve detailed description when discussing the environment of pneumatics. Energy, utilised for cooling and heating to establish a comfortable environment, consumes a considerable amount of available terrestrial resources. It seems futile that in order to extract energy from a room by cooling, energy must be consumed. This apparent futility further prompted Laing's investigations.

When solar radiation reaches the edge of the atmosphere small amounts, mainly in the blue region of the spectrum, are reflected, diffused and absorbed by the atmosphere; hence blue sky. Considerably more is reflected and absorbed by the clouds. Of about 50 per cent reaching the earth's surface, the majority penetrates the surface and about half of this is re-radiated as longwave terrestrial radiation. In addition, small energy transfers take place by convection, conduction, reflection and evaporation. However, the influence of the atmosphere and the ground on the heating effects of solar radiation is less important than is generally accepted. Alto-stratus clouds, the most frequent type in overcast skies, reduce the amount of solar radiation reaching the earth's surface by only about 35 per cent. Although heavy cumulus clouds reduce the amount very considerably, they are rare, and normally move rapidly across the sky, so that their influence is mostly negligible. This means that large amounts of solar energy are still available for man's utilisation, although these amounts decrease greatly when the sun's altitude is less than 20°. On the other hand, the influence of clouds, on energy losses due to longwave terrestrial radiation, is much more significant. The clouds act as a black body, absorbing this radiation and then transmitting more than half of it back towards the earth.

If heat energy is to be accumulated, conduction and convection heat exchanges must be minimised, whilst terrestrial radiation is prevented during the night. However, if on the other hand low temperatures are to be achieved, the prevention of solar radiation gain during daylight is essential. This heat build-up is, to some extent, observable in the single skin air supported structure, the 'greenhouse' effect. This effect could well be augmented by employing a wall system, transparent to solar radiation, yet reflecting the terrestrial radiation, and at the same time having high insulating properties. In this way large temperature increases could be achieved. These considerations lead to Laing's realisation of 'eternal day' and 'eternal night' periods.[4] For 'eternal day' energy is accumulated during the

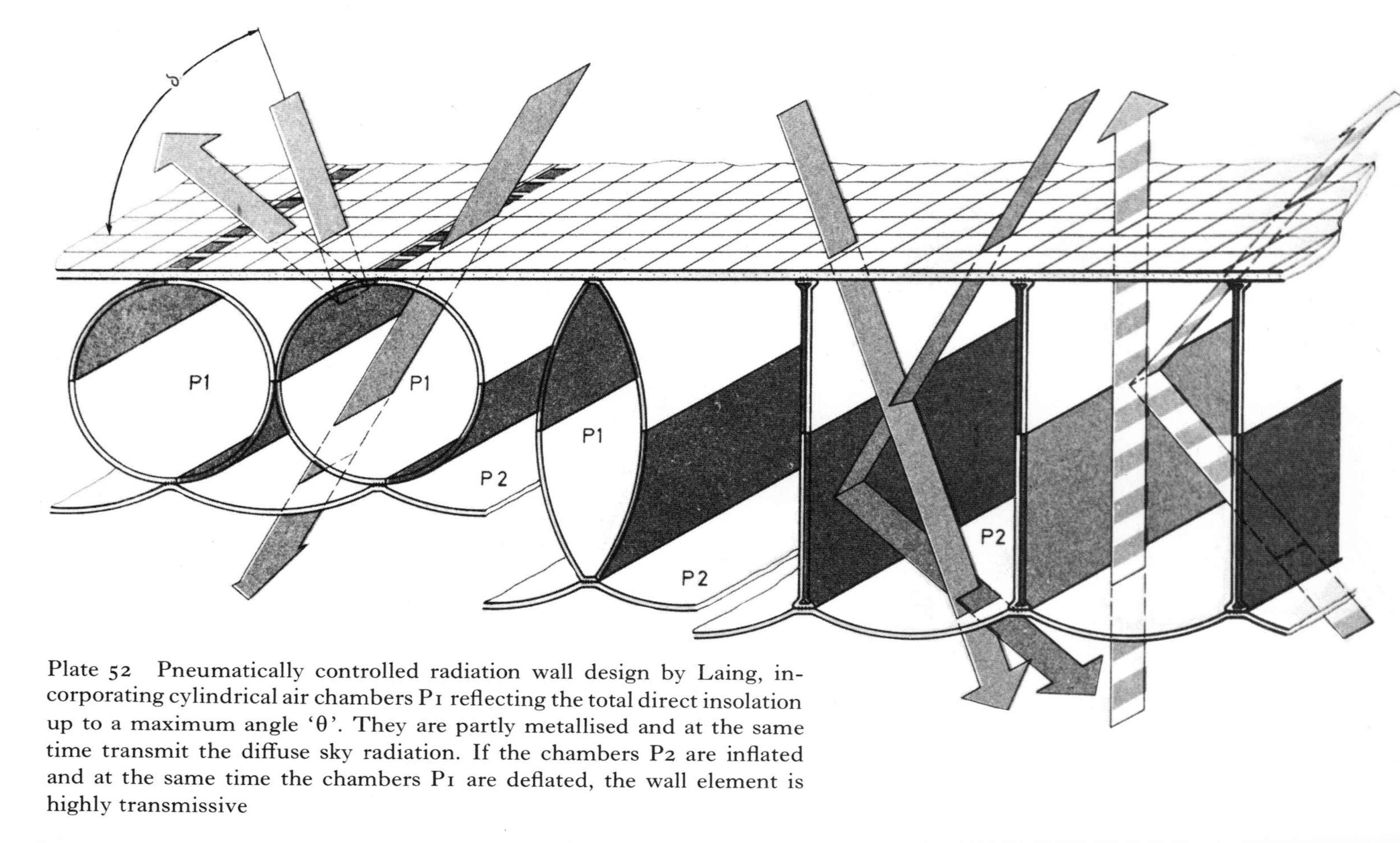

Plate 52 Pneumatically controlled radiation wall design by Laing, incorporating cylindrical air chambers P1 reflecting the total direct insolation up to a maximum angle 'θ'. They are partly metallised and at the same time transmit the diffuse sky radiation. If the chambers P2 are inflated and at the same time the chambers P1 are deflated, the wall element is highly transmissive

day, and is imprisoned by a reflective and insulative barrier during the period of no insolation. Further energy is accumulated during the following days, until saturation point is reached. As the temperature increases, less energy is accumulated during insolation periods, so after a while, the energy accumulations of the day and the small night energy losses are equalised, the point of saturation. In a similar manner, a temperature decrease, 'eternal night', can be achieved, but diffused atmospheric radiation reduces the magnitude of the temperature change.

In pneumatic buildings, the structural membrane has very little capacity for heat storage, and consequently only the floor or ground remains as a heat energy store. Its thermal properties are thus most significant; earth, water and concrete all absorb approximately 90 per cent of solar radiation and have good thermal capacities. The heat build up is radiated from the floor into the building interior during the night. If this heat radiation is not sufficient, greater thermal capacities can be achieved with circulating fluids that increase the heat transfer between the body of the floor and its surface. To achieve full control of terrestrial and solar radiation, complex wall configurations are essential, which in one instance are highly reflective, and in the other are highly transmissive. Energy transfer by conduction must be minimised, yet not at the expense of optical characteristics.

Laing investigated many different wall systems.[5] One of the simplest systems consists of transparent inflated sections containing partly metallised movable films, the positions of which are controlled pneumatically. In one position radiations are completely transmitted, even those arriving obliquely. In the other position, no radiation can penetrate the wall. In order to retain the brightness of the metal coating the inflation is with dry air or an inert gas, and these inflated sections provide heat insulation. With a double wall of this construction, Laing has obtained a heat insulating value corresponding to a 320 mm thick brick wall, the weight relationship of 1:1000 for the same insulation value being very impressive. With other wall designs by Laing direct insolation is rejected, yet diffused daylight freely penetrates the wall. These transparent films need to have definite optical properties appropriate to various climatic conditions. For cold climatic conditions where energy accumulations are called for, the films must be highly transparent to the ingoing short wave solar radiation. On the other hand in tropical climates, where cooling is essential, films with high transmissivity for the outgoing long wave terrestrial radiation are needed.

Plate 51

Plate 52

The great advantage of Laing's proposals is their dynamism. The transmission controllable walls can be continuously modified to suit climatic conditions and to achieve a specific environment. Not content with just the ability to control temperature, Laing sought for full air conditioning, which includes humidity and air flow factors.[6] By manipulation of the wall systems so that some elements are open and

Plate 53

others closed, local differences in air temperature can be created, and these will induce air movement within the building. On the other hand condensation can be produced by inducing long wave radiation emission from the walls, which causes cooling of the wall structure, and hence condensation. In this way the humidity of the air can be varied. This means that with these walls the climatic environment can be fully controlled within wide limits.

Despite the apparently complex nature of these walls Laing claims that their fabrication is quite simple, and that a continuous production method is viable.[7] Because of their relatively small thickness, compared with the overall dimensions of building structures, doubly curved sections can be quite easily fabricated. With mass production the cost of these wall units will be low compared with that of conventional construction. Also because of weight ratios in the order of 1:200, transportation costs will be greatly reduced. However the maintenance and control of such walls will naturally be higher than otherwise. A genuine comparison of costs is very difficult, but Laing suggests, that over a long period of time, this system could show savings of up to 70 per cent over conventional methods.[8] From his research Laing made the following conclusions:

> 'The proposed method liberates the inner climate from external climatic limitations and extends human habitat beyond the presently favoured regions. Uninhabited mountain regions, enjoying mostly clear weather, could be used for preferred construction sites, enjoying year round early summer climates.
>
> Solar energy controlled climates could diminish the current depletion of fossil fuel resources. It is possible that less developed countries should show cultural progress, when their adverse climate could be improved by solar radiation control, which could be the key to the development of their creative ability.'[9]

Lightweight pneumatic structures, in which the transmission of radiation is controlled in such a sophisticated manner as suggested by Laing are a complete contrast to the massive structures which are generally associated with either the utilisation of solar energy or the conservation of environmental energy produced by internal activities. Probably, it will be some time before Laing's ideas are fully realised, and undoubtedly present environmental control in pneumatic structuring on the whole falls far short of this standard of sophistication. Although this is so, at least Nikolaus Laing has made pneumatic designers aware of the environmental controlling possibilities in this new form of construction.

The Acoustic Environment

Acoustic conditions within pneumatic structures are far from satisfactory. In the first place, pneumatically generated structures are

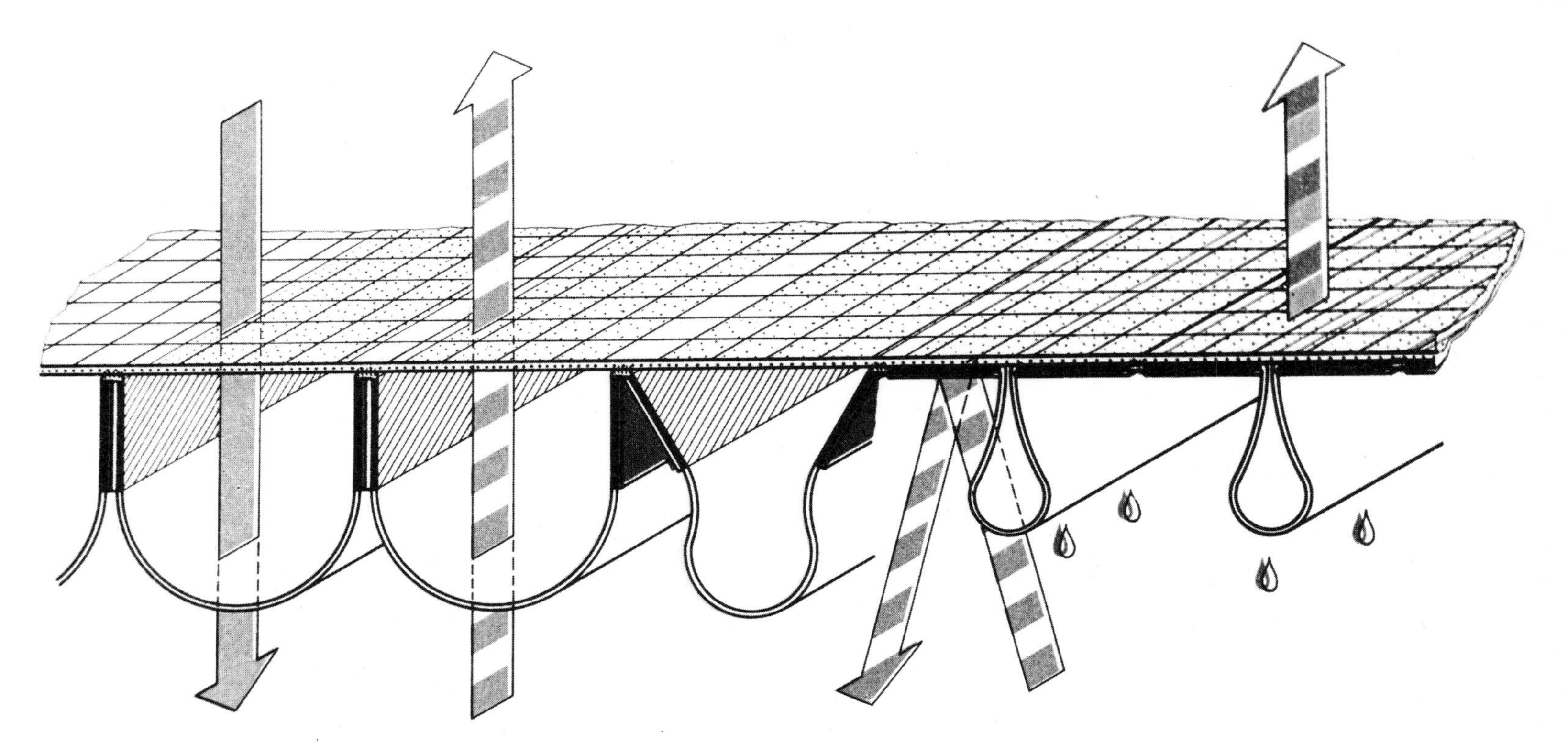

Plate 53 Pneumatically controlled radiation wall design by Laing which in 'closed' positions (to the right) turns a reflective surface to the inside and a black surface to the outside. The strong cooling effect of this wall during the night by ground radiation will induce the condensation of water inside the building

usually of a spherical or cylindrical form, which tend to focus reflected sound waves, causing echoes. For this reason, shallower structures are sometimes used, so that the focal point of the sound is below floor level. Although this gives some improvement, the acoustics still leave a good deal to be desired. However, with the use of cables, structures can be moulded into forms with favourable acoustic characteristics. With cable nets the smooth curved surfaces could be broken up still more, causing diffusion of sound waves, and providing much better acoustic differentiation. These suggestions appear feasible, especially considering the good acoustic characteristics of Lundy's A.E.C. exhibition building. In this, the interior roof configuration was moulded to prevent the reflection of sound waves to the centre of the exhibition area. In addition, the inside surface of the membrane could be lined with absorbent materials such as expanded polystyrene and glass fibre quilting.

Besides the problems caused by sound reflections within the structure, there is also the problem of sound transmission through the walls. The requisites for the reduction of sound transmission are mass and stiffness, and both these are totally opposed to the inherent characteristics of pneumatic structures. Marginal reductions are achieved with double membrane structures separated by large air spaces. Further reductions are possible by containing granular substances between membranes. Other alternatives are the use of leaded vinyl curtains, or the attachment of sound absorption materials to the membranes. However, the sound insulation characteristics of any lightweight structure have always caused severe problems. At the moment the only answer seems to be to tackle the problem at source and give careful consideration to acoustic zoning on the site. Obviously the acoustical engineering associated with pneumatic structuring needs much thought and investigation.

The Visual Environment

Since the main materials used for pneumatics are translucent, the amount of light passing through the membrane is sufficient for most purposes. The sunlight is uniformly diffused by the material itself, and is free of glare. However, on dull cloudy days supplementary light sources may be necessary. With transparent materials, such as mylar and transparent vinyl films, higher lighting standards can be achieved, but problems of glare may be introduced.

Few people have exploited the possibilities that these translucent and transparent membranes offer, particularly in terms of colour. The rich golden light, filtering through a white vinyl coated fabric membrane into the internal space, is an extreme contrast to the often dirty grey external appearance. Exciting and endless combinations of translucent, transparent and opaque materials, tied together with intricate jointing patterns, are possible, and these effects can be enjoyed internally during the day and externally at night.

Plate 54

Plate 54 Air supported radome at Bochum, West Germany, manufactured by Krupp, glowing in the evening light

Plate 55
Light sources are directed onto the membrane, which acts as reflector in this Gourock exhibition structure

Plate 56 Lighting on independent standards by Barracuda

Plate 57 Lighting units suspended from the membrane cause local stress concentrations in the fabric; a Gourock air supported structure used for food preparation by Robert Wilson and Sons Ltd

When artificial light sources are essential, utilisation of the structural membrane as a huge light reflector is an obvious solution, and

Plate 58 Fluorescent lighting tubes attached by fabric sleeves to the structural membrane of a Barracuda air supported structure

this helps emphasise the spatial qualities of the pneumatic form. Light fittings are mounted on the floor around the building's periphery, so that light is directed at the membrane reflector, and bounced back into the central floor area. With a white vinyl coated fabric membrane about 70 per cent of the light is reflected back, to illuminate the interior. Of the rest, 10 per cent is lost to the outside world and the other 20 per cent is absorbed by the membrane, and converted into heat. An alternative solution is the provision of special lighting standards, completely independent of the structure, which can act as supports for the structural membrane in the event of failure of the inflation equipment, thus preventing collapse of the structure. Quite often lighting equipment is suspended from the membrane material, but this is rather dubious practice on two accounts; firstly stress concentrations are introduced into the membrane, and secondly, movement of light fittings due to structural deformations can be quite disturbing. If this practice is adopted then it is essential that the lighting units, usually fluorescent tubes, be attached by fabric sleeves to the structural membrane. Maintenance of the high level equipment can be accomplished from step ladders, or on calm days the inflation equipment can be switched off until the lighting units and membrane fall to a workable level. By continuous control of the inflation equipment the units can be kept at this level indefinitely, but this must only be attempted on a calm day. On a windy day the membrane would flap about and the lighting equipment could quite easily be damaged; moreover the maintenance workers are in grave danger of being injured. If the structure is reinforced with cable networks then it is quite safe to attach lighting as well as other equipment directly to these networks.

Plate 55

Plate 56

Plate 57

Plate 58

Unfortunately pneumatic designers have been over-involved with the solving of structural problems. This structural enthusiasm has meant that environmental and aesthetic considerations have been sadly neglected. These latter considerations have usually received secondary attention, and in some instances were regarded as only worthy of an afterthought. A complete architecture, appropriate to present-day technology can only be achieved if every design aspect, whether functional, structural, environmental, sociological or aesthetical, is afforded equal consideration from the very first conception.

Maintenance

Many existing air supported structures appear to function efficiently despite a complete lack of maintenance by their owners. Regular checking of the building is essential if its safety is not to be impaired. This not only includes maintenance of the inflation equipment, but a thorough check of the membrane and anchorages for structural failures, and also the operation of entrances and exits. Maintenance of the inflation equipment is necessary and fan bearings must be oiled regularly. It is essential that maintenance includes the checking of any stand-by equipment. Anchorages need inspecting for soundness and the membrane for abrasive wear, and tears. Although inflation equipment can still support the membrane when it is quite badly torn, these failures must be repaired immediately since they can quite easily enlarge and place undue loading on the inflation equipment. Repairs of small tears that do not seriously affect the stability of the structure, can be accomplished *in situ* by fixing fabric patches over the hole with adhesive. Although this can be done with the building inflated, it is more easily achieved when the membrane is on the ground, fully deflated.

In industrial atmospheres the membrane soon becomes dirty, and therefore for aesthetic reasons cleaning of the membrane with a soap solution is advisable, just as windows on a conventional building are cleaned regularly.

Capital and Maintenance Costs

Although the cost of pneumatics can vary considerably, the figures for typical air supported construction prove extremely low. A floor area of 1000 m^2 can be covered for as little as £5000, £5 per square metre of floor area including membrane, anchorages, openings and inflation equipment, but excluding any foundations. Naturally with the more refined entrances such as air-curtains, this cost figure will increase. Since it is the accessories that constitute the more expensive items, and not the membrane material, the cost per square metre usually decreases as the size of the structure increases. Perhaps because the costs involved are negligible, very few studies have been made on the costs of erection and dismantling.* Assuming again a typical air

* During the winter of 1966/67, John Laing Research and Development

supported structure of 1000 m^2 floor area, erection will cost about £300 and dismantling about £75. Inflation fan running costs are about £3–£5 per 1000 m^2 of floor area per week. However, air curtains are rather expensive in power requirements.

A direct capital cost comparison with other forms of building construction can be very misleading. To make a true cost comparison not only must capital and maintenance costs be examined, but cost gains due to quicker utilisation of the building through greatly reduced erection times, cost gains due to the ability to utilise the site space for other purposes when the pneumatic is dismantled and not in use, cost related to the life-span of the building including depreciation, insurance costs and running costs must also be considered. Such a thorough costing breakdown has yet to be achieved; but these are likely to substantiate the low cost implied by the above figures. These figures emphasise the economy of this form of construction, so low a cost that a genuine throw-away product is implied and this is indeed appropriate in our age of planned obsolescence.

Scope and Limitations

Air supported construction has already established itself as one of the most efficient forms for portable and temporary enclosures, and undoubtedly for these applications its characteristics are unequalled by any other known form of construction.

Its most important advantage is without doubt its portability and mobility; the membrane material is deployed in the most efficient way possible, that is in tension; it is therefore extremely lightweight and is of convenient low bulk when packaged for handling and transportation; finally, erection and dismantling are accomplished swiftly and with ease. In addition to these, large unobstructed spans can be achieved and this allows great flexibility in the use of the building.

Considerable flexibility is also evident in the way the building volume enclosed by this type of construction can be varied quickly and easily by subtraction or addition of parts. This is certainly a distinct advantage in this present time of rapid change, particularly when it is combined with highly competitive construction costs.

Because of early failures people are still rather sceptical about its safety. However, the safety of this form of construction can be ensured with careful design. Quite apart from this, its safety factors are of a different kind from those of more conventional building forms. Structural failure in these latter forms is usually sudden and without warning, apart from deflection and the possible cracking of finishes, and if people are trapped within the building the consequences can be disastrous. With air supported construction the structure collapses slowly generally leaving the occupants ample time to escape, and even if they are trapped within the membrane they can feel their way out

---

Limited conducted site trials on a Barracuda air supported structure. Following these trials an internal report was presented in 1967.

from underneath, since the membrane is so lightweight. Admittedly equipment suspended from the membrane could be dangerous, but as yet in the whole history of air supported construction no fatal injuries to occupants have been recorded.

The other major hazard in building construction, apart from the structural collapse, is fire. Fire tests on air supported buildings have not been very conclusive. Although the coated fabrics generally used for the structural membrane do not support combustion, they melt when subjected to high temperature. This means that if a fire occurs within an air supported building, a hole will be melted in the structural membrane which will consequently collapse on top of the fire, causing the utter destruction of the membrane. How quickly this destruction is accomplished depends on the extent of the fire. Certainly ample escape exits must be provided to permit quick evacuation of the building.

Despite its many attractive characteristics, air supported construction is not without its limitations. The three problems unique to air supported construction have already been discussed at length, that is the need to maintain the pressure differential across the membrane with a constant air supply, the need to minimise air leakages, and the need to counteract the uplift forces with some means of anchorage. The former two have certainly presented problems, which, though surmountable, do place limitations particularly concerning access to and from the building.

Another major limitation of these thin membrane structures is their environmental characteristics, particularly their thermal and acoustic insulation. These can be overcome by the use of more complex constructions but this adds greatly to the cost.

Unfortunately these sensitive structures are susceptible to vandalism; they offer a challenge to the vandals 'to pop the bubble'. Extensive damage must be suffered before deflation occurs, and although this can usually be repaired quickly and easily, it is most inconvenient. It is also a temptation for vandals to try to deflate the structure by opening all the doors. Vandalism is often repelled with ugly barbed wire palisades; this not only revives the memories of concentration camps and prisoner-of-war camps, but also portrays a sorry reflection of our society. It is certainly reassuring, though somewhat surprising, to note that where no precautions have been taken against vandalism, very little trouble has occurred.

AIR INFLATED CONSTRUCTION

Low Pressure Inflated Rib Structures

Plate 59

In its simplest form air inflated construction consists of pressurised tubular framing, supporting a weather proof membrane. With small structures, low pressure ribs are suitable, although very large diameter low pressure tubes can cover large spans. Unfortunately these low pressure tubes are prone to diurnal temperature and atmospheric

Plate 59 R.F.D.-G.Q. Company's Numax low-pressure air inflated rib structure

Plate 60 Compensating air bags on the Numax shelter to minimise pressure variations

Plate 61 Frankenstein's high pressure air inflated rib cocoon structure

Plate 60

pressure variations, which affect the rigidity of the structure. For small structures integral compensating bags must be fitted to minimise these variations in the inflation pressure. However large structures need continuous automatic control of the pressure, and quite often need a continuous supply from a compressor, or even low pressure fans.

High Pressure Inflated Rib Structures

Plate 61

With the smaller high pressure tubes, these pressure variations do not affect the stability of the structure, but these high pressures require much higher performance standards for fabric material, jointing and sealing. Once inflated to pressures of around 700,000 N/m^2, the tubes retain this high level of pressurisation for considerable periods and only require topping up, as little as once every 3 months. Generally high pressure tubing is constructed from a circular woven sleeve of fabric, similar to the fabrics already discussed for air supported construction, with an inner airtight lining of extruded polyvinylchloride, neoprene or similar elastomer, and an outer weatherproof elastomer coating, hypalon being the most favoured. The ends of the tubes are sealed with metal fittings that are retained in position with circular metal clamps. Inflation is through a valve, very similar to a standard car valve, by means of a compressor. High safety factors must be used in this type of construction, since fabric failure can cause quite dangerous explosions. With tube diameters of $\frac{1}{2}$ m, spans of up to 20 m can be safely achieved. This inflated tubular construction allows quite a freedom of structural form. Many such structures do not make full benefit of the membrane covering, which if pretensioned between the tubular framework adds greatly to the structural efficiency of the whole structure.

Inflated Dual Walled Structures

The dual walled inflated structure is an obvious development of inflated tubular structures, basically being merely a row of adjacent tubes. These walls are held together by diaphragms or drop cords. 'Airmat' construction is suitable for smaller structures, thicknesses of 150 mm to 200 mm being pressurised at between 20,000 and 80,000 N/m^2 for stability. Usually this pressure need only be topped up occasionally. These forms of structure are inflated in minutes, with only small air inputs, and undoubtedly compressors are most suitable for accomplishing these inflations. Although 'Airmat' is extremely expensive, and also difficult to repair in the event of punctures, it can be fabricated into quite intricate structural forms. This is achieved by varying the lengths of the drop threads, so altering the contours of the membrane walls. It is generally only used in very specialised circumstances, since at the moment its expense rules it out for normal building applications.

Fig. 84

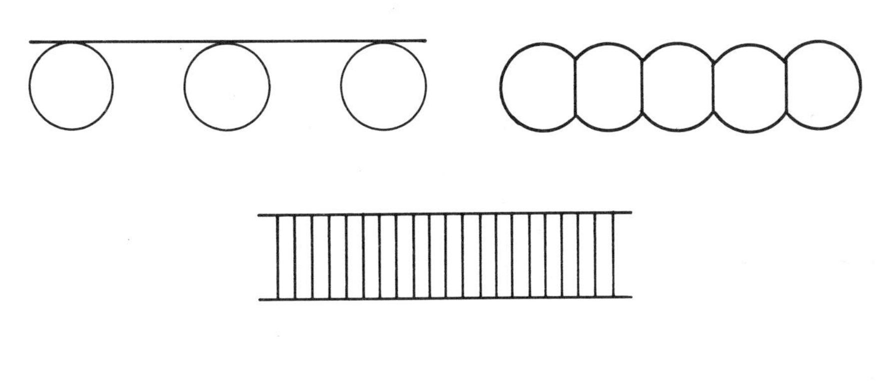

Fig. 84

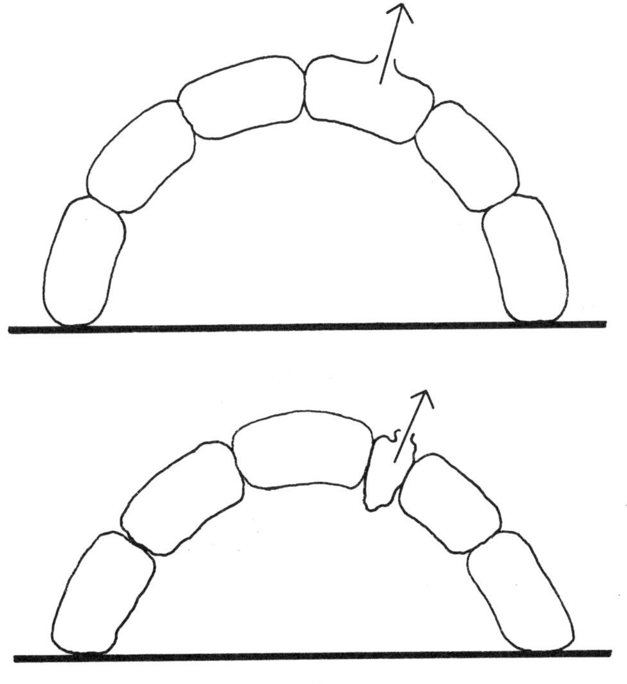

Fig. 85

Low pressure dual walled diaphragm construction can be used for quite large spans, but this necessitates thick wall sections. For the large volumes of air enclosed by this form of construction, a continuous air supply from low pressure compressors, or even fans, is appropriate.

Fig. 85

Air inflated construction can be easily adapted for compartmentation, so minimising the risk of structural collapse in the event of significant air leakages. In addition to this, these volumes of enclosed air have valuable insulation properties that can be well utilised for environmental control. At the present time connection of the dual walls is by simple diaphragms or drop threads, but the structural efficiency of the inflated elements can be greatly improved by more complex means of inter-connection, such as diagonal bracing, and indent construction; these have been studied carefully by Frei Otto.[10] Whether the cost of these more efficient constructions will be prohibitive, only the future will reveal.

Fig. 86

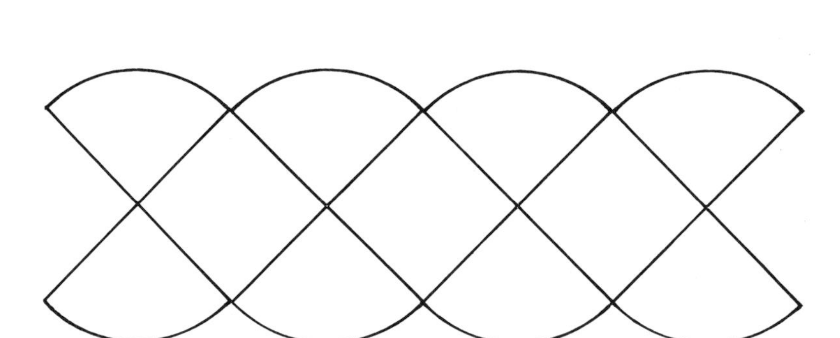

Fig. 86

Inflation Devices

Although in many other pneumatic applications sophisticated inflation devices have evolved, inflation methods for air inflated buildings are generally rather crude. With this form of construction inflation equipment must perform more than mere pressurisation of the structure. Firstly the building must be deployed and erected from its package state. Then it must be pressurised to its working inflation load. Because of air leakage and pressure variation due to temperature changes, the pressure level must be recharged from time to time. Finally if rapid deflation of the structure is required artificial aid may be needed. Quick inflation is always desirable, but caution is essential to avoid damage to the membrane fabric, especially to the diaphragms or drop threads, with high velocity air. Air inflated structures need some form of non-return valve through which inflation can be accomplished. For quick inflation and deflation these must be of compatible size to the structure; small valves of about 20 mm diameter are sufficient for small structures, but for larger structures valves of up to 200 mm are necessary. Fabric sleeves have proved to be unsatisfactory, not being very durable, so generally some kind of metal non-return valve is utilised. Some inflated structures may remain standing for quite long periods, and the enclosed air may suffer many compressions and decompressions due to temperature variation. If this air contains a high percentage of water vapour, then some of this is bound to evaporate. It is therefore advisable to protect the fabric with an internal elastomer coating.

For smaller structures the quickest form of inflation is undoubtedly from compressed gas or air cylinders and these are extensively used for inflatable survival equipment. For inflatable life-jackets inflation is by means of a small compressed carbon dioxide cylinder, activated by pulling a toggle which releases the gas into the jacket. Life-rafts are also inflated by this means although recent developments have produced an air aspirated system which achieves a saving in weight, very important for survival equipment carried on aircraft. This equipment comprises an aspirator unit, operating on the venturi principle, connected to a compressed air bottle. On activation the compressed air is released into the structure through a jet pipe contained within the aspirator unit. The flow of compressed air draws air in from the atmosphere so that up to five times the compressed air volume enters the structure. This not only achieves considerable savings in weight, but also provides a more rapid inflation system. Furthermore this system is considerably less affected by temperature variations than the carbon dioxide method of inflation. Subsequent recharging of the structure to maintain constant pressure can easily be accomplished, provided the compressed air cylinders have not exhausted themselves. If instant architecture is to become a reality, then similar inflation methods to these must be employed.

The most common method of inflation at present used is by blower or compressor. Whatever type of equipment is used the air supply equipment need only work at a pressure slightly above the required inflation pressure. In this way the equipment is working most efficiently, air not being greatly compressed by the equipment only to be decompressed as it enters the structural element.

With most equipment there is a reduction in the volume supplied as the pressure is increased; the 'Rootes' blower is an exception, maintaining a constant supply volume despite pressure increases. Automatic topping up to maintain the pressure level is easily achieved, provided that the equipment is maintained in good working order.

Deflation, however, is quite a problem. To evacuate all the air from an inflated element without artificial means takes some considerable time although collapse of the structure due to the weight and pressure is fairly rapid provided the valves are of compatible size with the structure. For packaging into the minimum volume the air must be completely evacuated, and this is best achieved with a decompressor. This equipment could considerably speed the dismantling of the structure and as deflation is controlled, the weather need not dictate when this can take place.

Where very large volumes of air are enclosed within structural elements, a continuous air supply is generally necessary because of the slight air permeability of membrane materials and jointing methods at present employed. In these cases the inflation equipment will be very similar to that used in air supported construction.

Scope and Limitations

Maintenance as regards air inflated construction is very similar to that for air supported construction, except the inflation equipment is not quite as important in most cases. It is very difficult to generalise about the cost of air inflated construction, since it varies so widely. Because of the double membrane, it can be said that the cost is at least twice as much as air supported construction, and with complex airmat construction it can be five or six times greater.

Air inflated construction is most suitable for rapid deployable structures, as is evident by its sole use for survival equipment. It does not normally present the problems of continuous air supply, air access associated with air supported construction, but its spanning potential is rather limited. Although the volumes of air enclosed within the structural elements have valuable insulation properties which can be utilised for environmental control, the double membrane increases the cost considerably over that of air supported construction. Like all pneumatic construction, portability, mobility and flexibility are its main attributes, and undoubtedly it is these facts that have popularised its use for so many different applications.

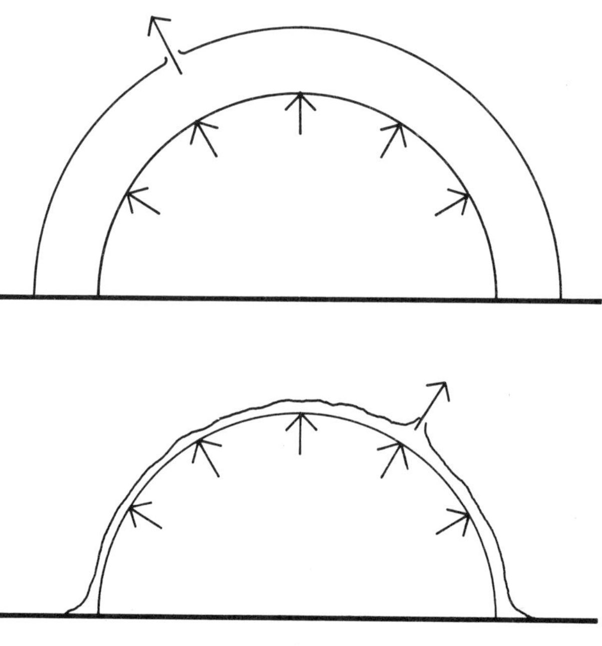

Fig. 87

HYBRID CONSTRUCTION

This form of construction has evolved because of the limitations of both air inflated and air supported construction. As defined in the introduction 'total pneumatic hybrids' integrate the features of these two constructional systems. The spanning potential of air supported construction is retained, insulation characteristics are increased, safety against structural collapse is improved, and the problems of access are reduced. In a similar manner pneumatic limitations are overcome with 'partial pneumatic hybrids' in which pneumatic construction is integrated with more conventional forms of construction. Undoubtedly the cost of this form of structure can vary extensively depending on the degree of constructional sophistication utilised. A few existing examples are evidence that it is in this form that sophisticated pneumatic building construction will achieve outstanding architectural heights. Koch's Boston Arts Centre Theatre and Lundy's exhibition pavilion for the United States Atomic Energy Commission have been previously described; other examples are mentioned in the final chapter.

Fig. 87

BUILDING REGULATIONS AND STANDARDS

In different countries throughout the world building regulations and standards vary widely, but the situation concerning pneumatic structures appears to be universally similar, in that legislative bodies attempt to apply existing building regulations. In many countries the building regulations are avoided, as pneumatics are classed as temporary structures. As far as the United Kingdom is concerned, building is controlled in two ways, by Town and Country Planning Acts and by the Building Regulations. In the former case permission is required for building if the structure is to be erected for longer than 28 days or if it is to be used as a place of public assembly. The official attitude on Building Regulations is mainly concerned with escape if collapse should occur. This necessitates the provision of a minimum width of emergency door openings, and these must be outward opening. The Ministry of Housing and Local Government and the Ministry of Public Building and Work* were aware of the need for some form of specification to control pneumatic building standards, and are at present investigating the situation in co-operation with pneumatic manufacturers. This confused situation appears to be universal, and if it continues much longer, could well be detrimental to pneumatic development. Some form of specification standard is highly desirable, but it must be realised that an adaptation of existing Building Regulations would be archaic for this entirely novel form of construction. This specification should cover the following points, which should be related to function and time factors:

---

* These two ministries are now part of the Department of the Environment.

1 Structural design of membrane and anchorage
2 Performance of materials and jointing methods
3 Design of inflation equipment as regards
   a Capacity based on air losses
   b Duplication or stand-by equipment
   c Ventilation requirements
4 Fire precautions as regards
   a Ample escape provision
   b Material incombustibility
5 Access provisions.

All designers conversant with pneumatics must strive to pass on their knowledge to official bodies, to encourage them and advise them of the required standards applicable to this type of construction. In this way, perhaps, Building Regulations can be devised which are appropriate to pneumatic behaviour, and which do not ignorantly inhibit the inherent potential of pneumatic structuring.

REFERENCES

1 Frei Otto in his book *Zugbeanspruchte Konstruktionen*, Band 1, p. 298 ff. examines anchorage methods in depth.
2 N. Laing, 'The Use of Solar and Sky Radiation for Air-Conditioning of Pneumatic Structures', *Proceedings of the 1st International Colloquium on Pneumatic Structures*, University of Stuttgart, 1967, p. 163 ff.
3 N. Laing, *op. cit.*, p. 163.
4 N. Laing, *op. cit.*, p. 166.
5 N. Laing, *op. cit.*, p. 169.
6 N. Laing, *op. cit.*, p. 172.
7 N. Laing, *op. cit.*, p. 174.
8 N. Laing, *op. cit.*, p. 175.
9 N. Laing, *op. cit.*, p. 176.
10 F. Otto, *op. cit.*

5. Para-Architectural and Non-Architectural Pneumatic Applications

Applications for pneumatics outside the bounds of architecture are so extensive, that a comprehensive study of all pneumatic applications is certainly beyond the scope of this book. However, certain non-architectural applications are relevant to this architectural discussion, since developments in these fields may be considerably advanced and may indicate future trends in architecture. Before examining these it would be as well to mention some of the pneumatic structures which appear more relevant to architecture, that is those which are used as auxiliary construction devices.

AUXILIARY CONSTRUCTION DEVICES

Pneumatic Formwork for Concrete Construction

Inflated membranes are economical minimum surface structures, which are ideal forms for concrete shell construction. In the early fifties inflatable membranes were used as models for such structures. Whereas the inflated membrane model is in tension only, the thus constructed concrete shell has only compressive stress acting upon it, along with some very small bending stresses in boundary regions. The logical development of this was to use an inflatable membrane as a concrete formwork. As long ago as 1942 Wallace Neff of Los Angeles started to experiment with pneumatic balloons, using them as formwork for concrete shell construction. Although he demonstrated that very cheap building structures could be constructed in this manner, he encountered technical difficulties, which prevented immediate utilisation of this method of construction. He sprayed concrete onto air supported domes, which were pressurised between 50 and 200 mm of water pressure. This pressure was not sufficient to prevent excessive deformation caused by the weight of the concrete during construction, which resulted in cracking of the concrete. To use high enough pressures that reduced deformation sufficiently to avoid this cracking, would have entailed bigger and more costly foundations, so Neff tried to limit the swelling of the form by using steel reinforcement rings. However, Neff's research did not uncover

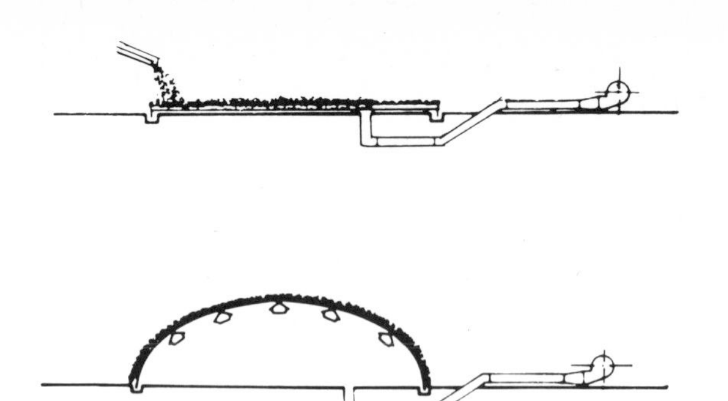

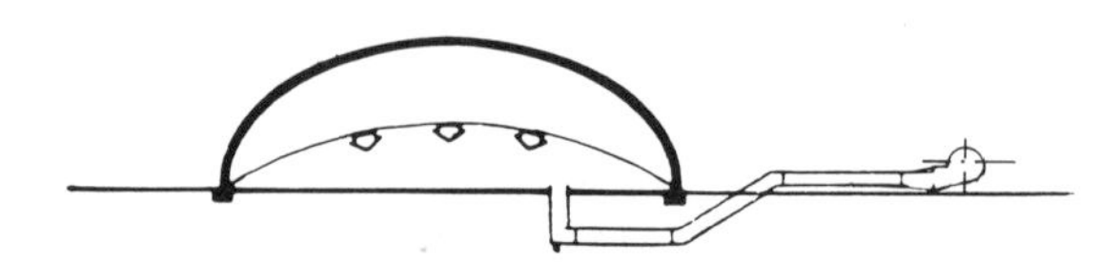

Plate 62 Construction sequence with the Binishell method for pneumatically formed concrete domes

a viable solution to these difficulties and hence the long time lapse before this form of construction was exploited commercially.

Binishells—Pneumatically-Formed Concrete Shells

Eventually during the sixties Dr. Dante Bini developed a viable system for the erection of concrete shells employing pneumatic formwork.[1] His patented construction technique uses an airtight, flexible neoprene membrane anchored to a peripheral foundation. An expandable steel mesh reinforcement is placed over the sheet and is enclosed by a much thinner plastic sheet which as well as containing *Plate 62* the concrete during erection provides a colourful weather-proofing to the dome. The entire mass is raised into position by inflating the lower membrane to a pressure of between 200 and 600 mm of water pressure. The steel mesh as well as reinforcing the concrete, prevents the mix from sliding off. It can be either a mesh of steel springs that can adapt themselves to a variety of final shapes, or an articulated form of mesh that restrains the dome to a particular shape. The concrete is vibrated immediately after inflation, and because of its thinness cures in a short time allowing the neoprene membrane *Plate 63* formwork to be struck after between 12 and 48 hours. Insulation can be adhered to the inner surface, or even a second concrete skin can be constructed, allowing insulation and services to be positioned be-

Plate 63 A typical Binishell concrete dome

tween the two skins. Openings for such things as doors, windows and services are simply cut in the concrete with a circular saw. The viability of this form of construction has been substantially proved by numerous examples in Italy for barns, warehouses, offices, houses and sports centres, which have undergone extensive structural tests. The great advantage of these concrete domes lies in their ease of erection, where conventional formwork, scaffolding and cranes are eliminated, and in their strength. However, only shallow domes can be constructed, heavy foundations are needed as a counterweight against the rather high inflation pressures used, and the complex steel reinforcement is rather expensive.

Low-Cost Concrete Forms by Heifetz

An Israeli architect, Haim Heifetz, has recently made a study of the use of pneumatic formwork and has overcome the problems of cracking due to deformation of the formwork encountered by Neff and has developed a system which he claims is considerably more economic than the Binishell. Instead of placing the concrete at ground level, it is sprayed on to an already inflated balloon, pressurised as high as 1000 mm of water pressure. These high inflation pressures can be counteracted by radially arranged demountable trusses, or by a reinforced concrete floor slab. In the former method

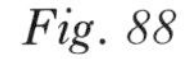

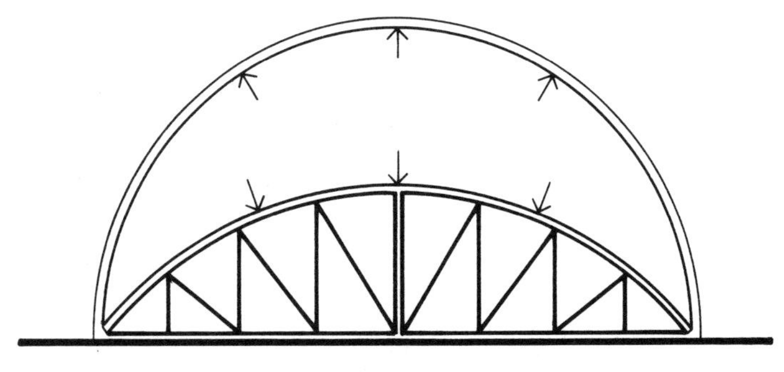

Fig. 88

Plate 64 Frankenstein's pneumatically formed, rigidised foam structure

the balloon formwork does not come into contact with the ground, preventing the inflation pressures from being transferred into the ground and this means that only minimal foundations are needed. Using this method, many dome shaped houses have been built in Israel at a cost of about £15 per square metre of floor area, construction being completed in less than a day.

Pneumatic Forms for Plastic Dome Constructions

With the development of plastics it might well be that the spraying of plastics on a pneumatic form will hold greater promise. Furthermore, rigidising foams can be introduced into inflated structural elements to form permanent rigid buildings.

Frankenstein, a British firm, has developed a method of rapid building construction especially suitable for applications in remote areas such as desert, jungle or polar regions. A liquid is sprayed on to both the inside and outside of an air supported form, and immediately turns into a low density rigid form. The building, which can be constructed in 3 to 4 hours, consists of the pretensioned membrane sandwiched between the rigid foam 100 mm thick on the inside and 20 mm thick on the outside. As the foam is carried in the form of dense liquids of low bulk, the materials for this form of building can easily be transported by air. This transportability, along with its quick erection, lends this form of construction to the housing of the homeless, particularly in disaster areas, and to exploration and military work, but as yet it has been very little utilised.

Plate 64

However, as recently as April 1970, a similar form of foam plastic construction was used to good effect in a disaster area. Many towns and villages in Western Turkey were devastated by an earthquake

Plate 65 Bayer's rigidised foam plastics homes provided temporary accommodation for earthquake victims in Turkey

leaving many thousands of people homeless. Bayer, a German company, immediately flew some of their engineers into the stricken areas accompanied by a transportable spraying unit, auxiliary plant units and polyurethane raw materials. Like the Frankenstein system, the constructional process is very simple. The polyurethane foam is sprayed on to the inflatable form as it rotates on a turntable. A 5 m diameter by 3 m high dome shaped shelter can be sprayed up within the hour. Openings can then be cut out for windows and doors, and ground insulation can also be provided. A total of 400 of these homes, each housing eight to ten people, were erected and provided very effective shelter for many refugees.

Plate 65

Despite the use of pneumatic formwork for building construction beginning over 30 years ago, it is still exploited very rarely. Only Bini and Heifetz can truly be said to have developed their methods for concrete construction sufficiently for practical application and only Bayer has proved the viability of foam plastic construction. Low cost make it suitable for a wide range of usage, but with its quick erection times, undoubtedly its major application could be in overcoming the mammoth problem of the homeless.

Pneumatic Lifting Devices for Building Construction

Besides applications for concrete and plastic formwork, pneumatic devices are already often used in the building construction process, and recent developments have shown that they could be of even greater value. Pneumatically controlled tools, like the pneumatic drills, are familiar construction tools. Pneumatic lifting devices, although as yet little used for building construction, have proved very effective for aircraft or vehicle salvage work. Pneumatic lifting bags

Plate 66

Plate 66 R.F.D.-G.Q. Company's pneumatic lifting bags, used for aircraft salvage

provide a speedy aid to the salvage of crashed or disabled aircraft and are particularly useful for recovery from soft, uneven or marshy ground, where conventional jacking methods are impractical. On a smaller scale upturned vehicles can easily be righted, and inflation can be accomplished using the vehicle's exhaust system. In the U.S.A. inflatable bags have been used to raise the roof of single storey prefabricated buildings into position, and on a much larger scale, giant balloons have helped in the erection of geodesic domes. These pneumatic lifting devices obviate the need for expensive cranes, and often scaffolding is not needed, but their great advantage is that extremely heavy structures can be quite easily raised.

At the Ministry of Technology Research and Development Establishment, Cardington, England, an inflatable hoist has been developed. It consists of a length of hose which is clamped between a pair of rollers; on inflation the rollers are forced along the hose, and this movement can be used for lifting loads vertically or hauling vehicles horizontally. At the moment this inflatable hoist is being used for the inspection and maintenance of electric pylons. A line is fixed over a suitable attachment point by rocket and a 70 mm diameter hose is hoisted in position. A pressure of 20,000 N/m^2 provided by a small compressor is sufficient to lift an average size man up for maintenance purposes.

Plate 67

Plate 67 A pneumatic hoist, capable of lifting a man

Airships

To many people the days of the airship are long since passed mainly because of the frightening disasters that eventually ended their commercial use in 1937. However, it is generally agreed that this system of transport was developed before technology was sufficiently advanced to solve all the design problems that emerged. Recently there has been increasing suggestions for the return of the airship.* The

* M. Rynish, an aviation writer, has recently made a special study on the airship with reference to the building industry, but his ideas have been viewed with scepticism by former airship designers.

Plate 68 A mobile pneumatic structure, a hovercraft developed at Cardington Research and Development Establishment, where air not only provides the motive force but also the structural formwork

validity of the airship for future uses is strengthened by the possible need for a new and more efficient transport system. Whereas land and sea transport can only serve individually a limited area of the earth's surface, air transport can cover directly every inch of it. The airship could carry large loads for long distances, picking up its cargo and setting it down anywhere. This means that building construction would not be restricted by present-day land transportation routes, which were inherited from by-gone ages and that now largely dictate urban development. Complete fabrication of buildings could then be a reality, a house being finished in the factory down to the last paint-work, lifted *in toto*, and carried directly to the site, where only the services need be connected. Mass production techniques more akin to car production than the present-day building methods, would speed urban development as well as reducing costs. Such a transportation system would also have striking implications on civil engineering projects. Large bridges could be lowered directly into position, huge building equipment could be moved easily and quickly between sites, and raw materials could be carried in bulk, relieving the roads of the many fleets of lorries associated with the building industry. Although the benefits of such a transport system appear immense, the setting up of an entirely new transportation system would be very costly. Support would be essential if such a system was to be implemented, but it would indeed be a bold Government who sanctioned such a transportation upheaval.

Plate 69 Pressure lifting vessel developed at Cardington Research and Development Establishment

Ground Effects Machines

With the invention of the hovercraft or ground effects machine, the first truly amphibious vehicle, that exhibited compromise to neither land nor water, was born. The air cushion lifts the vehicle above the land and the water, and thus frees movement and communications from the hitherto fixed transportation routes of road and rail. It is difficult to forecast whether such an air cushion device will form the vehicular basis of a future completely mobile home, but it must be remembered that these devices are still at an early stage of development.

Hovercraft have received much publicity both in the popular press and technical literature, so much so that it would be pointless to discuss them in detail here, although one craft is worthy of mention. This, an inflatable hovercraft, which can be packed into a very small space for storage, or for transportation by land, sea and air, was developed by the Research and Development Establishment at Cardington, England. The craft, 6 m long by nearly 3 m wide, is big enough to carry four people, and can be packed into a size 1 m^2 by $\frac{1}{2}$ m deep, small enough to be carried in a car boot—perhaps a very crude prototype for a pneumatic mobile home.

Plate 68

This principle of supporting loads on an air cushion is of great use for the transportation of heavy equipment. With wheeled vehicles heavy loads are not easily carried across soft terrain and besides this, existing bridge constructions are often not capable of supporting heavy concentrated loads. Air cushion devices attached to these

Plate 70 Inside the pressure lifting vessel

wheeled vehicles, merely distribute the weight of the vehicle evenly over a larger area preventing it from sinking into the ground or lessening the loading effect on bridges.

Cardington Research and Development Establishment have progressed one stage further, by designing a pressure lifting vessel which can travel over water. This vessel consists of an inflated tubular base and a fabric canopy fitted with a number of suspension attachment points. The centre part of the base can be deflated so that the base can be hinged to allow the vehicle to drive into the vessel. The vehicle is attached to the canopy with suspension wires. The canopy, when inflated by a low pressure blower, lifts the vehicle off the ground and enables one man to push the vessel, complete with vehicle, easily over land or even over water. Such a vessel could considerably ease the movement of equipment and materials on a building site.

Plate 69

Plate 70

Smaller air cushion devices, hoverpads, are used for the conveyance of goods in industry. Air cushion grass cutters are also becoming popular, since they work just as efficiently on steep slopes. Air cushion beds have been developed for the hospital treatment of severe burn cases; the very small pressures exerted on the tender skin cause very little if any discomfort. These are just a few examples of air controlled structures, which have as yet been little exploited.

Plate 71 Inflatable survival life-raft, sophisticated instant structuring

PORTABLE PNEUMATIC EQUIPMENT

Inflatable Survival and Escape Equipment

Such equipment seems far removed from the world of architecture, but it is relevant because in no other pneumatic structure is such sophistication evident. Research into survival equipment has been abundant over the last 30 years, and though equipment is now very advanced, further developments are constantly being marketed. Materials have to withstand extreme conditions of exposure, but more important, inflation devices must be 100 per cent reliable. This highly developed equipment sets standards of sophistication that pneumatic building designers should seek to achieve.

Both inflatable life jackets and survival life rafts are prime examples of pneumatic structures, which are highly compact, when stowed and not in use, yet are instantly operational when inflated. Inflatable escape slides have proved to be the most efficient means of evacuating disaster-stricken aircraft, and can also be designed to function as a life-raft. A considerably more sophisticated escape device has been developed by the Goodyear Aerospace Corporation; an air inflated stairway, known as the 'Inflatostair', it also can be inflated in seconds to permit evacuation from danger areas such as stranded aircraft or burning buildings. It is the construction of these stairs that is most

Plate 71
Plate 72
Plate 73
Plate 74

Plate 72 Container in which the inflatable life-raft is stored

interesting. This is of 'airmat' type construction utilising dropthreads of varying length to connect the dual walls. In this way complex structural configurations are possible, which illustrate the versatility of 'airmat' construction.

All the above mentioned inflatable equipment, whether for escape or survival can all be classed as instant structuring. Their degree of sophistication and their efficiency are both ample evidence that sophisticated pneumatic architecture can be achieved and that instant pneumatic architecture is indeed a reality, and not just a utopian dream.

Pressurised Storage Containers

Although animal skins are still used only by a few primitive civilisations for liquid storage, pressurised membrane containers for both

Plate 73 Quick evacuation by an inflatable escape slide

liquids and solids are becoming more and more popular. Such containers differ slightly from pneumatic membranes stressed by air pressures in that the weight of liquid or solid tends to deform the container to a flattened form similar to a liquid drop. Pressurised membrane containers for water storage and carriage are already frequently seen on camping sites. Encouraged possibly by the success of these containers, several companies have developed huge collapsible fabric tanks, which are now used to store virtually any liquid, from drinking water to petroleum products. Tanks of a million litres capacity can be deflated into a package of size $1 \times 1 \times 5$ m and are ideal for liquid storage in remote areas. Inflatable dams have also been developed in close conjunction with these tanks. These are generally secured to a concrete base and abutment, and are inflated with water or air at low pressure. The height of the dam can be varied

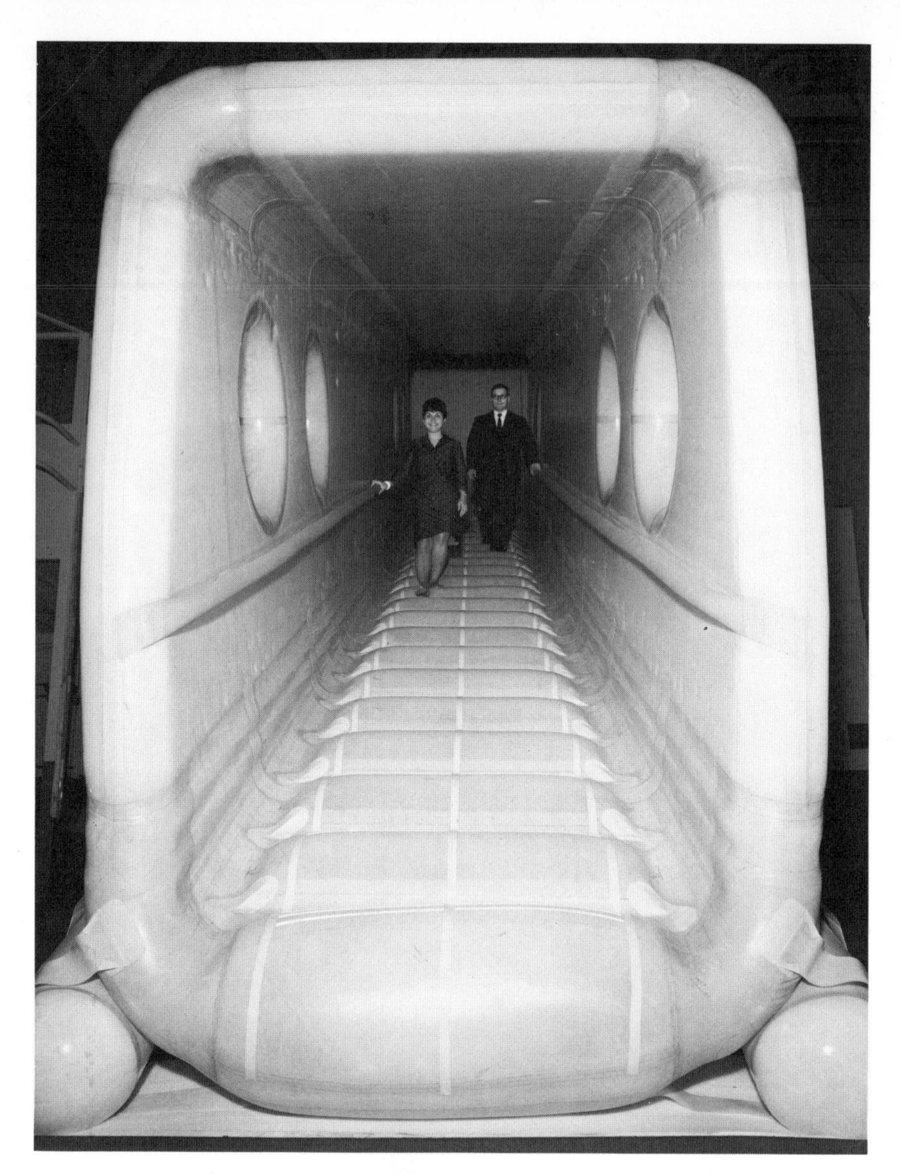

Plate 74 An air inflated emergency escapeway developed by the Goodyear Aerospace Corporation

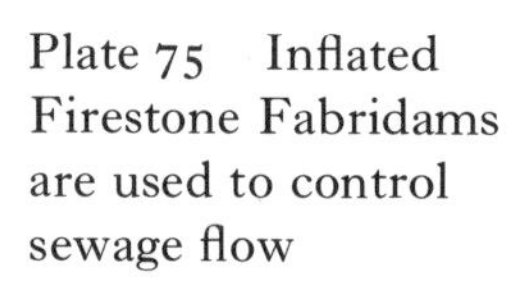

Plate 75 Inflated Firestone Fabridams are used to control sewage flow

Plate 76 Firestone Fabridams, inflated with air or water, have been frequently used in America

by further inflation or deflation. 'Fabridams' manufactured by Firestone have been tested satisfactorily for heights of up 7 m; for handling convenience they are generally not longer than 150 m, but can be joined together by intermediate concrete abutments. Smaller, but similar 'Fabridams' have also been used for sewage control.

Plate 75

Plate 76

Pneumatics for Leisure Pursuits

Leisure pursuits have been responsible for a wide range of pneumatic applications, and indeed these have done much to familiarise people with this form of construction. Beach balls and inflatable buoyancy rings bring back youthful memories. But one of the greatest exploita-

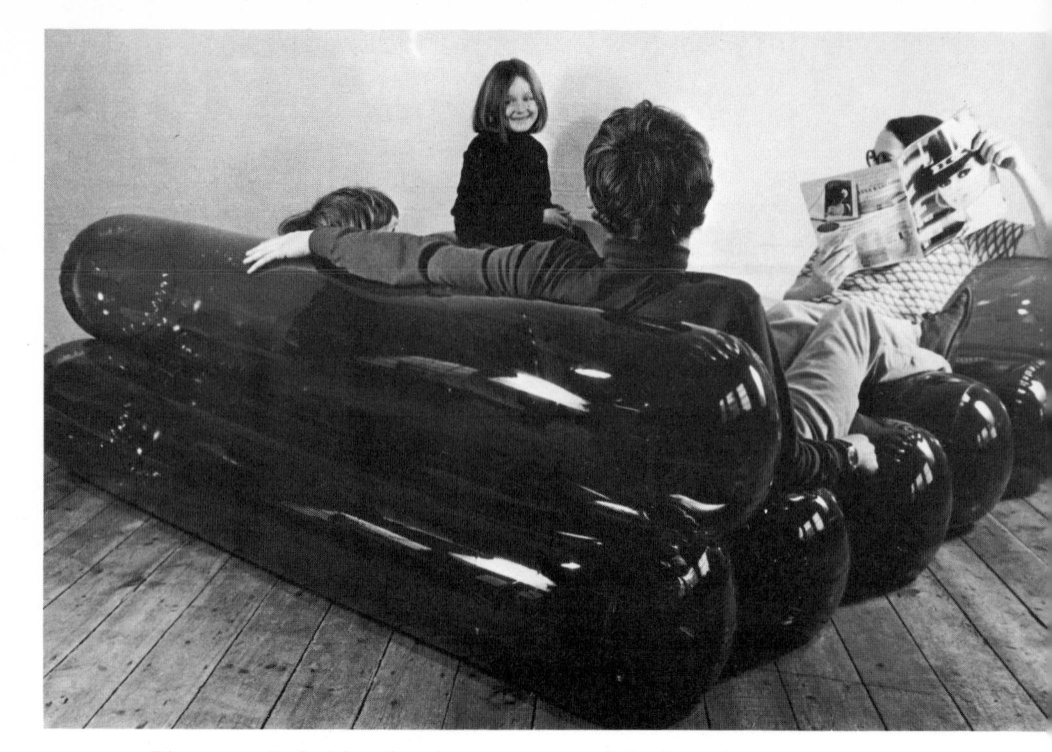

Plate 77 Inflatable furniture, a craze of the late sixties

tions of pneumatics has been and still is for camping equipment; in fact the very first pneumatic shelter was the Stromeyer inflatable rib tent of pre-Second World War days. Few people would dream of going camping without their air inflatable pillow and mattress. Air inflatable furniture is a frequent sight on beaches and camping sites the whole world over. Its use is now no longer restricted to camping; during the late sixties inflatable furniture came into vogue for every-day use. The pioneers of this craze were undoubtedly the French architects Aubert, Jungman and Stinco and the designer Quasar Kahn. Their work, besides bringing inflatable items into the forefront of furniture fashion, has done much to familiarise the public with pneumatics, so opening the doors to pneumatic architectural applications.

Plate 77

THE POLAR EXTREMITIES AND BEYOND: TWO CASE HISTORIES

Exploration Logistics

One of man's natural instincts is to explore and further his knowledge of the unknown. Now that virtually the whole of the earth's surface is within his grasp he is now trying to thrust himself into the realms of outer space. In all these exploration ventures into remote areas, whether it be to conquer high mountains, to penetrate the jungles, to search for natural wealth in polar and desert regions, or even to plummet into space, equipment portability and mobility is essential. These requirements certainly validate an examination of pneumatic suitability for exploration purposes. As far as earthbound exploration is concerned, polar regions still present man with his greatest challenge and are therefore worthy of close study.

Logistics of Polar Exploration

Very little research work has ever been undertaken on the economics and logistics of providing building structures for polar exploration, but investigations by the National Research Council of Canada into transportation costs to the extreme Northern Territories, indicate, that if traditional building techniques were used in polar regions, the cost of transporting labour and materials would be far greater than the material costs.[2] Materials are normally transported by sea to the polar regions during the summer months, and are subsequently distributed within the continents by air and land. Although shipping costs may appear low against land and air transportation costs, even they could add about a third to the original cost of a typical, yet conventional, portable shelter.

This high cost of transportation immediately suggests that local materials should be used wherever possible. However, the only local materials of the polar continents are ice and compacted snow, and these are of limited value for building. Because of this lack of suitable building materials, construction materials must be transported to these regions, and consequently lightweight materials which have a high strength to weight ratio must be employed in the most efficient way possible. As well as this, buildings must be quickly erected, to minimise on labour needs, and also because of the frequent heavy storms, which can destroy partially completed buildings, or at the best, greatly delay further building operations. These requirements appear to be well satisfied by pneumatics, but as yet there are very few examples of pneumatic utilisation.

Building Design Criteria in Polar Regions

The major influences which affect building design in these regions are as follows[3]:

1 Inaccessibility
2 Lack of energy resources
3 Adverse climate:
 a Low temperature
 b High wind speeds
 c Snow accumulation
 d Snow drifting
4 Snow properties
5 Fire hazards.

With maximum wind velocities in excess of 200 k.p.h., conditions frequently occur in which outside work is impossible. Although lower temperatures are experienced inland, coastal storms are generally more severe and more frequent. Very little snow actually falls in the vicinity of the poles, but annual snow depth increases as great as 1·5 m are encountered near coasts.

Polar regions are plagued with drifting snow which accentuates the problem caused by natural snow accumulation. Initially, this problem was met passively, by allowing snow to accumulate over and around building structures until the contours of the structures were smoothed out, so that very little further drifting occurred. However, great loads were superimposed on the structure, ventilation was curtailed and access hampered. Consequently, great efforts must be made to minimise or even eliminate the effect of drifts; and this can be accomplished by either streamlining of structures, planning of the structural layout to produce wind velocity increases which enable deposition of snow beyond the buildings, regular maintenance and rigid discipline concerning the positioning of surface obstructions.

Quite extensive investigations on snow and ice properties have been undertaken, so that it is now possible to predict the behaviour of any structures built on snow fairly accurately.[4] These structures will undergo a settlement, the magnitude of which is dependent upon, firstly the snow characteristics, secondly the weight of the structures, thirdly the loss of snow strength due to the heat radiating from the structures, and fourthly the shape and size of the structures.

Inaccessibility of the polar continents has rendered the hazard of fire very significant. The lack of public fire fighting services and sufficient water for extinguishing fires at remote polar bases means that a fire outbreak could completely destroy a base, leaving the occupants at the tender mercies of the adverse climate. Consequently, design against fire is of the utmost importance.

Polar Exploration Bases

To enable a systematic taming of the alien environment of the polar regions several forms of community bases are needed. Permanent bases are essential, to maintain the required support for the varied activities undertaken within these continents. These activities consist of regular scientific observations, work in association with space exploration, technical and industrial experiments, and trans-polar expeditions. These bases can be either inland or coastal, but the latter afford the greater support. With the former, perennial snow is the major problem, and experience has proved the superiority of underground bases, the most notable of which are Camp Century, Greenland and Byrd Station and South Pole Station in Antarctica. Certain activities require temporary bases for short-term occupation, either for the establishment of permanent bases or for research work in remote areas. Such bases must be quickly built, and are normally established and supported by air. Portable bases are essential for land exploration. As with temporary bases the design criteria of these are,

Plate 78 Air supported igloo in use at Camp Century, Greenland

portability, and ease and swiftness of the erection and dismantling procedure. In addition to these, emergency bases may be required in the event of unforeseeable disasters, such as air crashes, fire destruction and isolation of exploration parties. For such bases some form of inflatable survival equipment, similar to that used at sea, but more extensive, to cope with the extreme environment, may be suitable.

Design Criteria for Portable Polar Shelters

In polar exploration, one of the greatest needs is for an extremely portable shelter. Such shelters are needed in areas where, usually, aircraft represent the most efficient means of transport available. Hence one of the most important requirements for these shelters is that their packaged bulk and weight is a minimum. Not only must living accommodation and storage space be provided, but shelters are also required for the efficient maintenance and repair of mechanical equipment. For such shelters the following general criteria are applicable:

1 Satisfactory operation in temperatures of minus 50° C.
2 Structural stability in winds of 160–200 k.p.h.
3 Transportability by air or tractor trains
4 Minimum maintenance requirements
5 Expected life of 2 to 5 years
6 Quickly and easily erected and dismantled
7 Simple in design
8 Fire resistant.

Plate 79 Numax shelter erected and in use as a workshop at Shackleton Base, Antarctica, January 1957

Portable Polar Shelters

For these shelters, generally of small span, air inflated construction appears to be valid. The pressurising requirements of air supported construction, at the moment, place too heavy a burden on fuel resources. However, with the portable nuclear power generators of the future, this may not be a problem. In actual fact, an air supported structure has been used at Camp Century, Greenland, where a small nuclear reactor provides energy, but this structure was mainly used for the construction of underground snow tunnels.[5] Insulation requirements also imply use of double membrane air inflated structures. The suitability of air inflated construction for polar use is exemplified by the performance of an inflated structure, used on the 1958 Trans-Antarctic Expedition as a vehicle maintenance shelter. This shelter was a Numax hut, manufactured by the R.F.D.-G.Q. Company, England, comprising an inflatable low pressure framework of tough rubberised fabric with a proofed-nylon cover and floor.

Plate 78

'. . . our engineers had used this temporary shelter of an inflatable rubber garage. This folded into the shape of a large valise and weighed about 120 pounds (55 kilograms). With a small electric blower it could be inflated in three or four minutes and then stood 30 feet long (9·150 metres), 15 feet wide (4·575 metres) and 9 feet high (2·750 metres), and could be picked up by four men quite easily. In conditions of changing temperature it had one serious and uncanny characteristic. When the sun was

Plate 79

> high its warmth heated the air within and the structure stood up firm and rigid, but as the sun sank, it shrank and slowly collapsed, enveloping as in a shroud, any vehicle it contained. Next morning in warmer conditions, it would again be fully erect, ready for the engineers to continue their tasks. It served us well until really low temperatures occurred when the rubberised cloth became brittle and torn.'[6]

Apart from its structural instability and poor low temperature durability, this shelter amply satisfied the general criteria applicable to portable shelters. Recent developments have since overcome both these drawbacks. The structural instability was caused by inflation pressure variations, due to temperature fluctuations. With low pressure inflated structures, these variations can be minimised by use of integral compensating bags, or by the automatic operation of the blower, when the pressure falls below the required level. On the other hand high pressure inflated structures are not unduly affected by these temperature variations. Correctly designed inflated structures, when sufficiently anchored, will resist winds of up to 160 k.p.h., with only small deflections, and these are generally not prohibitive. With regard to the low temperature durability of the 'Numax' shelter, it should be pointed out that this structure was not specifically designed for polar regions, and thus it is not surprising that the rubberised fabric failed. Recent developments have since produced materials with satisfactory low temperature performance.

However, the most frequently used material for pneumatic structures, vinyl coated nylon, does not appear suitable. Although minus 50° C. cold flexibility can be achieved with this material, both the adhesion of the vinyl film to the nylon and the fire resistance are reduced. Neither of these relaxations are acceptable in polar regions, the former making the material susceptible to damage by sudden impact. However, dacron fabrics coated with such elastomers as hypalon, neoprene, silicone and butyl possess satisfactory low temperature durability. In the field of plastic films, aluminium and polytetrafluorethylene have admirable characteristics, the latter remaining flexible at temperatures of minus 70° C., and possessing excellent weathering properties.

Portability Criteria

The degree of portability of any shelter is influenced by the following criteria:

1 The package volume per square metre of floor area, 'V/A'
2 The weight per square metre of floor area, 'W/A'
3 The erection time
4 The ease of erection
5 The durability under continuous utilisation.

Plate 80 ML air inflated dual walled structure erected in the snows of Norway

Compared to similar sized conventional maintenance shelters, the 'Numax' shelter undoubtedly has better portability characteristics.[7] The ratios 'W/A' of 3·4 and 'V/A' of 0·02 are as much as 10 times lower than comparable shelters, and the erection time, which is nearly instantaneous, is 300 times quicker. On top of this, its cost is between a half and a third that of conventional metal framed structures. These facts underline the potential of air inflatable ribbed structures for use in polar regions as shelters for stores and equipment, but unfortunately, their poor insulation characteristics make them unsuitable for living accommodation.

Portable Polar Living Accommodation

For the shelter of man, a much more sophisticated building structure is required. Such a structure is the dual walled inflated structure which, due to the air trapped between its two walls, offers superior insulation. Two such structures have been developed for use in polar regions and although they fulfil the general criteria applicable to portable shelters, they have yet to be truly evaluated in these extreme conditions. They are the ML inflatable shelter and the MUST self-contained unit.

The ML inflatable shelter is a half cylindrical structure with spherical ends, 16 m long by 9 m wide. Its dual walls are of 'airmat' type construction 125 mm thick. Each shelter is made up of four sections, which are joined together by zip fasteners, enabling the structure to be extended as required. The structure is inflated to 3 m of water pressure by an electrically driven compressor, which auto-

Plate 80

Plate 81 The smallest ML shelter, the 'igloo' type, 9 m in diameter, packed ready for moving

matically maintains this pressure level with occasional topping up. The details of the shelter are as follows:

Plate 81

Floor area	144 m^2
Weight including compressor	820 kg
Package Volume including compressor	12 m^3
'W/A'	5·7
'V/A'	0·08
Erection time	12 man hours

Tests have demonstrated the structural stability of the shelter in winds of up to 150 k.p.h. and also its ability to withstand extremely cold temperatures of minus 50° C. Besides satisfying all general criteria, applicable to portable shelters, the thermal insulation provided by its walls is outstanding. Unfortunately, much heat is lost through the floor, which is only heavy duty vinyl coated nylon. These heat losses can melt the snow ground covering, possibly causing heavy settlement of the structure. Insulated timber floors are available, but these are not very portable. Perhaps a better solution would be an inflatable floor, that had an upper rigid floor surface such as plywood or reinforced plastic. In this way high local loads could be easily supported and little portability is sacrificed. Lighting and air conditioning units can also be supplied by the manufacturers, but these increase the weight of the shelter to 1800 kg and the 'W/A' value to 12·5. Nevertheless this is still considerably better than more conventional portable shelters. Since this shelter is still in the early stages of development, its high expense, between £60 and £80 per

Plate 82 The MUST self contained unit

square metre of floor area fully equipped, should not detract from its otherwise excellent performance which was commented on in a report by the British Antarctic Survey.[8]

The MUST self-contained unit, developed by the Garrett Corporation in the USA, is an even more sophisticated shelter, which is described in full in the following chapter. The details of this shelter are as follows:

Plate 82

Total floor area	390 m^2
Total weight	9000 kg
Packaged volume	44 m^3
'W/A'	23
'V/A'	0·11
Erection time	20 man hours

Although not as expensive as the M.L. shelter, its cost, about £50 per square metre of floor area, is still high, but like all other pneumatic structures it possesses extremely good portability characteristics.

But how do pneumatics perform with reference to settlement, anchorage and snow accumulation? Pneumatic shelters are so lightweight that they barely exert any pressure on the snow. Provided the floor is sufficiently insulated, the heat generated within the building will not decrease the load bearing capacity of the snow significantly. Thus, in comparison with the more conventional building techniques, at present employed in polar regions, pneumatic shelters will have less tendency to sink in the snow. Compaction of the snow site, prior to erection of the shelter, increases the load bearing capacity of the snow, and thus minimises this settlement even further.

Pneumatic shelters have very little mass to counteract aerodynamic lift-off forces of the wind, and hence must be anchored to maintain their stability. Although snow ballast is plentiful, positive anchorages are preferable, and in polar regions these are normally achieved by 'dead men'. These consist of holes, dug into snow, which are filled with compacted snow after the anchor has been inserted.

Fig. 89

Pneumatic profiles are very streamlined and tend to lessen the accumulation of snow due to drifting. Even so, snowdrifts are still a major problem. Frequent maintenance and strict discipline is essential to keep the structure serviceable. In less severe climates, the small snow accumulations upon pneumatic structures are easily dissipated, but in polar regions, structures can become buried in a very short time, and be subjected to heavy snow loads. For short-term structures these problems are surmountable, but true assessment of the performance of pneumatic structures as regards snow accumulation cannot be made without extensive trials in the polar continents. Despite these reservations the great advantage in portability afforded by pneumatic structures will surely lead to their eventual use in polar regions. For garaging of vehicles and for storage, an inflated frame structure offers a cheap and satisfactory shelter; the MUST self-contained unit indicates the sort of highly sophisticated accommodation unit that may in future be used. Such a unit could be mounted on an exploration vehicle, or even be self propelled on its own air cushion, and at the touch of a button, it would pneumatically expand affording ample shelter from the severe climate. Sir Vivian

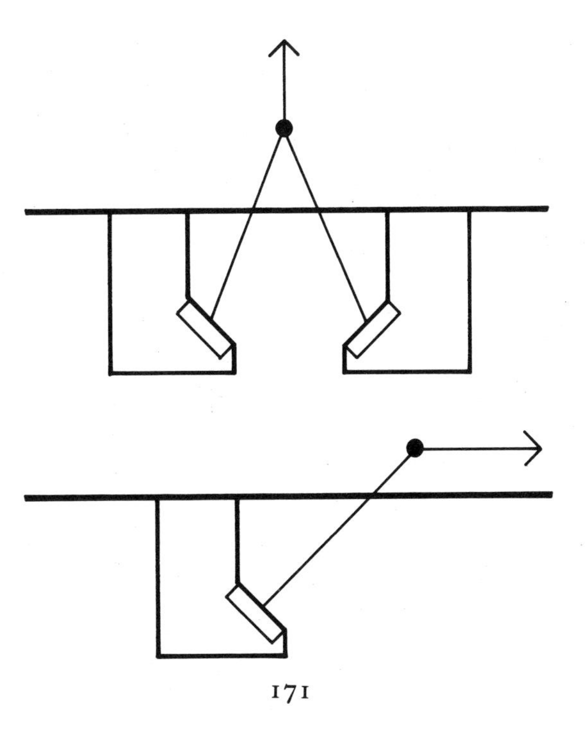

Fig. 89

Fuchs, that great authority on the polar environment, is of the opinion that there are potential uses for inflated structures as workshops, laboratories and living accommodation.

Temporary Polar Shelters

Temporary and portable shelters are of a very similar nature, except that whereas portable shelters are re-located at several sites, temporary shelters are erected at one site until more permanent structures are built, or at sites, where the period of use does not warrant a permanent base. Additional criteria are applicable to temporary shelters; they need not be reusable, but may be left unattended for short periods, and this means that pneumatic structures that need occasional 'topping up' of the internal pressure may be suspect. Although such topping up can be accomplished automatically, reliability of automated equipment cannot be fully guaranteed in the adverse polar climate. This reason is not sufficient to cause rejection of inflatables as a usable proposition, since present and future material developments will surely make inflatables completely air-tight, obviating the need for occasional 'topping up'. To combat this present disadvantage, air inflated structures could be rigidised by the introduction of rigidising foams into the structure, or foam plastic could be sprayed onto pneumatic formwork. Such a structure has also possibilities for a more permanent form of shelter.

Pneumatics at Permanent Polar Bases

In the past, fuel oil for heating, equipment operation and power generation has been the largest single logistic problem in the polar regions. However, this problem has been relieved at some of the larger permanent bases by the introduction of small nuclear power plants, in place of diesel generators. The great advantage of these nuclear reactors is that their fuel requirements are far less bulky. A 50 kg weight of uranium produces more power than 2 million litres of diesel oil. This means that at bases, where such plants are installed, energy is in abundant supply and therefore the power requirements of air supported structures are no longer a limitation on their use. It is interesting to note that before the introduction of nuclear power plants at these bases, diesel oil was stored in 45,000 litre flexible pneumatic neoprene tanks. These tanks, which weigh about 350 kg and may be folded into a package volume of 3·5 m^3, have been very successful.

Owing to the problem of perennial snow accumulation, under-snow structures are now used at inland bases for permanent shelters. In general these consist of ice caves in which independent light-weight prefabricated shelters are erected. Tunnels are constructed in the snow by the digging of trenches which are then enclosed by the compaction of snow over a suitable formwork. Usually this formwork has consisted of corrugated steel arches, but at Camp Century,

Greenland, an air supported structure was used as the formwork with great success. However, apart from this application, there appears to be little use for pneumatics at these inland bases.

Since the snow accumulation at coastal bases is generally not perennial, under-snow structures cannot be constructed, and underground structures would not be feasible because of the high construction costs implied by the frequent occurrence of permanent ground freezing. Hence, surface structures must be designed, that come to terms with the problem of drifting snow accumulation. At these more accessible bases, the problem of materials transportation is not as acute as at inland bases, and hence, conventional building methods do not have such severe limitations. However, great logistic savings could be achieved by using pneumatic structures. This is particularly true of large span structures, such as aircraft hangars, for which conventional building methods still have severe limitations.

There have been many utopian suggestions that whole communities in the polar regions could be enclosed within an immense structure, in which the environment is completely controllable.

> 'Professor Andrews noted that Frobisher Bay, Northwest Territories, had underground tunnels connecting its buildings, but he suggested that self-enclosed bubble cities might be more agreeable for inhabitants. . . .
>
> The bubble's design would have to offer minimum resistance to constant Arctic winds, he said, but at the same time catch and hold the snow for its insulating properties.'[9]

At the moment air supported construction presents the most feasible means of achieving such structures and indeed its use for this purpose has already been suggested.[10]

For small span structures, the most attractive building technique would appear to be rigid plastic foam construction, achieved with pneumatic formwork, but considerable research and trials into this technique are still necessary before it can be used for permanent shelters.

For large span structures, air supported structures may perhaps be the best solution in terms of material and transportation economics. The limitations of air supported structures, which were mentioned with regard to portable and temporary bases, are no longer applicable. Since nuclear power is generally utilised at these bases, the burden placed on the power supply by the pressurisation equipment is no longer limiting. At the moment, the available materials have a life of between 5 and 10 years, but this can be extended by re-coating the material, and soon, the present enthusiasm of material manufacturers will no doubt produce more durable materials. Sophisticated environmental control methods as suggested by Nickolaus Laing will indeed be appropriate for these applications. Air supported structures

Plate 83 Echo I, the world's first passive communications satellite

have already proved their potential for use in adverse climates. Since 1948 the majority of DEW-line radar stations along the extreme Northern Frontier of America have been housed within air supported radomes of about 15 m diameter. Few new techniques have been introduced into the building industry with such exacting climatic trials as these radomes, which have proved the potential of air supported construction for polar use beyond dispute.

Logistics of Outer Space Exploration

The high cost of transportation to the extreme polar regions are insignificant when compared with the probable costs of space transportation and labour. Estimated costs for placing a kilogram of material or equipment in orbit around the earth have been put at £1000, whilst to land the same on the moon will probably cost between £3000 and £6000.[11] Labour costs on the moon will be in the region of £30,000 to £50,000 an hour.[12] It can be seen from these approximate figures, that any materials or equipment, used in space, must have the maximum performance/weight ratio, and that only energy consuming activities, which are highly productive, can be afforded in space exploration.

Pneumatic Space Structures

Pneumatic structures possess many of the attributes which are essential for any space structure; these are high structural efficiency, low ratio of packaged volume to deployed volume, ease of erection,

Plate 84 Echo A-12 passive communication satellite, a rigidised balloon and its launching capsule

re-usability, portability and resilience. As early as August 1960 the U.S.A. launched a three stage Delta 1 rocket from Cape Kennedy containing a capsule only 700 mm in diameter in its nose cone. Once in orbit this capsule released a balloon, Echo 1, which inflated itself to a structure 30 m in diameter, quite a phenomenal explosion! The balloon, constructed of a thin mylar film 0·127 mm thick and covered with a vapour deposit of aluminium, was designed as a passive communications satellite, and only therefore required a surface capable of reflecting radio waves. The balloon also contained a mixture of benzoic acid and anthraquinone which continuously gave off a gas, so keeping the balloon fully inflated. This was one of the first pointers to the future role of pneumatics in space exploration. Soon after it was followed into orbit by Echo A-12, a rigidised balloon 41 m in diameter, which during launching was contained in a small spheroid only 1 m in diameter. Tests carried out with both these balloons before launching indicated that much research was needed to define

Plate 83

Plate 84

Plate 85 Inflated balloons and buoyancy rings ensure a safe splash-down for the Apollo astronauts returning from the moon

more clearly the dynamic forces occurring during the inflation process. However, Echo I was so successful, that a similar but larger balloon, Echo II has been manufactured and this has now been launched to join Echo I in a lower orbit around the earth.

Pneumatics have been featured extensively in both science fiction comics and novels for quite some time, and recent space research indicates that these fictional ideas are not far from the truth. On each of the United States Apollo moon missions three balloons covered millions of kilometres packed in shallow trays, ready to play a dramatic role in the final splash-down in the sea. Their solitary purpose was to upright the command module should it settle in the water upside down. When this happened, as it did on the Apollo 7 mission, the balloons were inflated by air compressors, and within minutes they had flipped the module right side up. Besides these balloons, the module was fitted with an inflatable buoyancy collar, which ensured that the module did not sink. Inflatable space stations, inflatable solar energy collectors, inflatable paragliders for re-entry deceleration of space vehicles and many other ideas are subjects of

Plate 85

intensive research work by the National Aeronautics and Space Administration. Pneumatic space structures are indeed things of the future.

Design Criteria for Lunar Shelters

A closer examination of the possible use of pneumatics for creating liveable environments on the lunar surface is most appropriate, remembering the successful moon landing of the United States astronauts Neil Armstrong and Edwin Aldrin in July 1969. Soon manned bases will be established on the lunar surface, and already prototype inflatable structures have been built. The design criteria for these base structures are as follows[13]:

1 Function of shelter
2 Use of high strength to weight ratio materials
3 Portability
   a Minimum packaged volume
   b Ease, quickness and reliability of erection
   c Ease of handling packaged structure
4 Meteor protection
5 Pressurisation safety
6 External temperature variation from minus 100° C. to plus 200° C.
7 Shielding against cosmic and solar radiation
8 External vacuum environment producing high-pressure loading of membrane, about 5 to 10 m of water pressure
9 Structural life of 3 to 5 years
10 Reduced gravitation on the moon, about a sixth that of the earth
11 Compatibility of structure with lunar surface materials.

Pneumatic Lunar Shelters

Because of the destructive forces present in the lunar environment, double structures appear to be a logical solution, that is a structure within a structure. A relatively thin outer structure could achieve sufficient protection from meteors, radiation and the extreme temperature variations. The inner structure would be completely independent of the outer skin, with the sole purpose of providing a pressurised internal environment. However, greater economy and a lighter overall structure could be obtained by incorporating this outer and inner structure into a single structural system, utilising a double or multi-walled structural concept. By variation of the distance between walls and the use of impact absorbing material, satisfactory protection could be achieved. Suitable coatings on the outer wall provide the necessary thermal protection, and access is more easily achieved with these multi-walled constructions. The 'airmat' has already been employed for many terrestrial applications and will probably be used for future lunar applications. Triple walls with staggered joints would give even more protection, and compart-

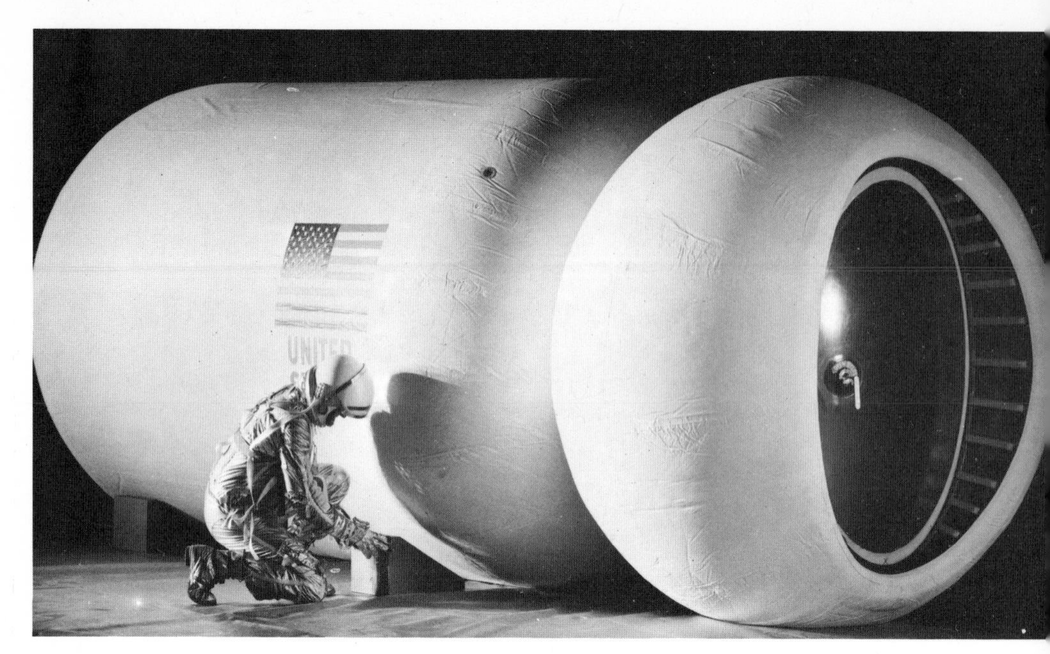

Plate 86 The STEM lunar shelter developed by the Goodyear Aerospace Corporation

mentation would eliminate the danger of punctures causing the depressurisation of the interior. The life supporting system, as well as including carbon dioxide removal, oxygen supply, food supply and waste disposal, requires the storage and circulation of large amounts of water. This water could be stored inside the wall compartments thus providing the means of pressurising these compartments. Besides this, such a system would have many other advantages; it would have a built-in puncture warning device, moistened surfaces being easily detectable; and as well as providing good protection against heat and radiation, water has an excellent thermal capacity which could be utilised during the 'lunar night'. Although such a constructional technique would be very heavy for terrestrial applications, the lunar gravitational force, only one-sixth of that of the earth, makes this technique more appropriate for lunar structures. Rigidised foams could also be used within the compartments after inflation. Such foams as polyurethane expand up to twenty times their original volume, but since the foam becomes quite rigid, re-usability as a portable shelter is sacrified.

The unique design criteria for lunar shelters demand the use of very specialised materials. The most important characteristic of these materials is that they should possess a high strength to weight ratio; gas impermeability, radiation resistance, good performance under large temperature variations, ease of fabrication flexibility, incombustibility and low cost are also essential attributes. Extensive research on the development of new and suitable materials for space and lunar applications is at the present being conducted, and this has already shown mylar and certain materials composed of minute whiskers, such as carbon, to have great promise.*

Present knowledge indicates that pneumatic construction is without doubt one of the most efficient forms of structuring that man knows for lunar applications. Already the Goodyear Aerospace Corporation has developed a cylindrical shelter 4 m long and 2·1 m in diameter for a NASA contract. Known as the Stay Time Extension Module, it is designed for use on post-Apollo program moon missions, and would support two astronauts for 8 days. Fabricated from flexible high strength stainless steel filaments in a composite with other flexible materials, it has a fully conditioned and self-supporting environment, maintaining a constant temperature of 21° C. Having been stowed in a canister attached to its lunar transport vehicle, the moon house would be inflated, after landing, by compressed oxygen. Once inflated, the shelter could be constantly manned by one astronaut, who could maintain communications with his team-mate exploring the lunar surface, with the orbiting command module, and with mission control back on earth. The Goodyear Aerospace Corporation claims that the prime advantage of the inflatable STEM concept is that it could be transported to the lunar surface as a small package and then be inflated to provide a habitable shelter; also its outer wall makes it highly resistant to meteoroids, thermal radiation and temperature extremes likely to be faced in the hostile environment of the moon. Certainly this will be the first of many pneumatic proposals for lunar applications, and it will not be long before at least some of these are realised.

Plate 86

Space 'Spin Offs'

The much higher standards of reliability, essential in space work, require large economic resources, which many feel are better utilised on more urgent terrestrial problems. However, 'spin offs' from space research are already influencing terrestrial structures. The sophisticated environmental control methods, essential for space exploration, will soon render obsolete many earthbound designs. Pneumatic suitability for these applications will help encourage the use of sophisticated pneumatic environments for terrestrial architecture, which are appropriate to our present-day technology.

* Development in space research is so swift that the written words get out-of-date overnight. Literature from the National Aeronautic and Space Administration provide the most up-to-date information.

REFERENCES

1. D. Bini, 'A New Pneumatic Technique for the construction of Thin Shells', *Proceedings of the 1st International Colloquium on Pneumatic Structures*, University of Stuttgart, 1967, p. 52 ff.

2. H. B. Dickens and R. E. Legget, *Building in Northern Canada*, National Research Council of Canada, Technical Paper No. 62, March 1959.
3. In a dissertation 'Pneumatic Structures in Architecture, with special reference to Arctic and Lunar Applications', School of Architecture, University of Liverpool, 1968, the author discusses these influences in greater detail. Further information can also be found in 'Symposium on Antarctic Logistics', *Proceedings of conference at Boulder, Colorado, 13–17 August, 1962*, Washington 1962.
4. 'Ice and Snow; properties, processes, and applications', *Proceedings of a conference held at Massachusetts Institute of Technology, 12–16 February, 1962*, Cambridge, Massachusetts 1963.
5. C. M. Daugherty, *City under the Ice, the story of Camp Century*, p. 86.
6. Sir V. E. Fuchs and Sir E. Hilary, *The Crossing of Antarctica*, p. 119, by kind permission of Cassell and Co. Ltd.
7. G. E. Sherwood, *Pioneer Polar Structures, Portable Maintenance Shelter*, Technical Report R-317, U.S. Naval Civil Engineering Laboratory, Port Hueneme, California, 1964.
8. J. R. Green, *Report on the ML Inflatable Shelter and its Potential in the Antarctic*, British Antarctic Survey, London, 1964.
9. *Toronto to teach Arctic Building*, The Polar Times No. 57. December 1963, The American Polar Society, p. 26.
10. F. Otto, *Zugbeanspruchte Konstruktionen*, Band I, p. 37.
11. *Materials for Space Operation*, Paper SP-27, National Aeronautic and Space Administration, Washington, 1962, p. 39.
12. '2000 A.D.', *Architectural Design*, February 1967, p. 67.
13. R. Szilard, 'Pneumatic Structures for Lunar Bases', *Proceedings of the 1st International Colloquium on Pneumatic Structures*, University of Stuttgart, 1967, p. 35.

6. Pneumatic Architectural Achievements

At first pneumatic building construction gained its popularity through its very novelty, but as people became aware of its advantage for certain applications over more conventional building forms, this popularity increased due to its efficient functionalism. The circumstances in which it warrants architectural consideration in place of traditional building methods are indeed extensive and grow even more as days go by. At the mention of 'pneumatics' the words 'temporary enclosure' spring to mind, but this term may be somewhat unfortunate. The word temporary is often associated with inferior products that are not worthy of serious design consideration. Until people realise that temporary products can have just as much, and in some cases, more validity than more permanent structures, then pneumatics will hardly attract the attention their potential deserves. Despite this association with the word 'temporary', the greatest attribute of pneumatics is their portability, which can be matched by very few other structural types. Quickly erected and dismantled, they can be packed into a very small volume for transportation. Besides this they have a flexibility far greater than most structures: their volume can be expanded or contracted almost instantaneously. As yet it is difficult to put any limit on the spanning potential of pneumatics, but it is certain that this is far greater than any other known structural device. All these attributes can be achieved at a fraction of the cost of any other means, and so it is obvious that there is a very wide field for pneumatic application. Up to now this has spread rapidly into military, commercial and social fields, but has achieved its greatest recognition in exhibition structures. The following pages look into these different applications in more detail, and in some cases make an appraisal of a specific building which it is felt demonstrates the implications of pneumatics in architecture far beyond mere enclosure devices and brings out their great versatility and potential.

MILITARY APPLICATIONS

Many new technical developments are initiated for military applications, and this is certainly true with pneumatics. The first air

Plate 87 An inflatable bridge developed at the Military Vehicles and Engineering Establishment, Christchurch, England

Plate 88 Inflatable radar antennae are widely used by the U.S. armed forces

supported structures were the DEW-line radomes, which have been previously mentioned. The excellent performance of these original radomes, under severe climatic extremes, proved conclusively the practicality of pneumatic building construction.

Since these early post-war days, this type of construction has found many widespread military applications both for engineering and

Plates 89, 90 The inflatable MUST shelter, a self-contained unit developed by the Garrett Corporation

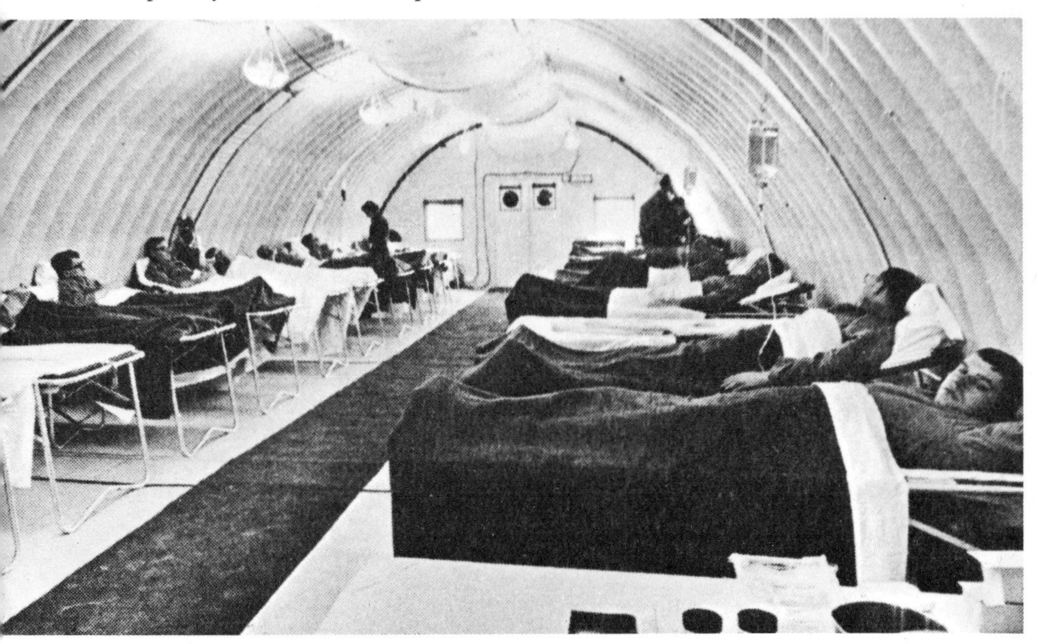

building construction. In the former these have ranged from inflated bridges to inflated radar antennae, whilst in the latter air supported and air inflated shelters are used for such things as storage, weapon assembly, vehicle and aircraft maintenance, hangars, garages, protection of radar equipment, mobile hospitals, portable living accommodation, to name but a few.

Plate 87
Plate 88

One particular military shelter deserves special mention because of its high sophistication. The MUST shelter, medical unit self-

Plate 89 contained transportable, is a mobile hospital unit that has been developed by the Garrett Corporation in the U.S.A., and as its name suggests is completely self-contained apart from fuel requirements. Basically, it consists of a utility element that forms the 'heart' of a pneumatically expandable complex. One utility element provides all the power for four inflatable shelters to operate at temperatures as low as minus 50° C. This element is 1·8 m wide by 2·7 m long by 2·4 m high and is equipped with a gas turbine to provide all the necessary power for heating, lighting and inflation. Each inflatable shelter is a dual walled semi-cylindrical structure, inflated to 10,000 N/m^2, 15 m long and 6 m wide, although sectionalisation permits any desired length to be assembled. A rigid integral insulated floor is provided, which is undoubtedly the bulkiest element in this otherwise highly sophisticated building structure. Unfortunately, such sophistication can only be achieved at the high cost of about £50 per square metre of floor area. Despite this, its extremely good portable characteristics, as well as its sophistication make it eminently suitable as a mobile hospital for disaster relief and civil defence besides its many military applications.

Plate 90

This is only one example of how intense design and research work for military purposes produces the ultimate in mobilisation. Design of military equipment tends towards miniaturisation, functionalism and portability, and these are indeed characteristics of pneumatic construction. Military design strictly adheres to clearly defined limits and within this design the function is never in doubt. Such design work provides a valuable stimulation for architects, who so often become 'bogged down' by their aesthetic partialities and ignore the fact that architecture is an art bounded by functional parameters.

COMMERCIAL APPLICATIONS

Whenever equipment has undergone satisfactory vigorous testing for military applications it is soon commercially exploited, and pneumatic building construction was no exception. Once the DEW-line radomes had established the practicality of pneumatics, commercial applications followed, and certainly the most rewarding has proved to be industrial.

Storage and Warehousing

Enclosed storage requirements have for a long time presented quite a problem to industrial concerns, for they can vary considerably from day to day. These fluctuating demands render permanent forms of building impractical. Admittedly warehouse space can be easily rented, but besides being expensive, this accommodation is quite often inconveniently situated, thus causing inefficiency through rehandling of stocks. Air supported construction has undoubtedly proved to be the answer to this industrial dilemma, in that it provides an instantly variable building enclosure volume. Companies unanimously claim

Plate 91

Plate 91 A Barracuda air supported shelter used as a temporary warehouse

Plates 92, 92a An insulated air supported structure used for storing fruit, exterior and interior views

the economic viability of this solution, but are reluctant to substantiate these claims with precise costing studies. Their reluctance is understandable, since it is extremely difficult to make direct comparison with other solutions. Such studies must consider, besides capital and maintenance costs, cost savings due to quick structural deployment and efficient space utilisation.

Although for many warehousing applications no control of the environment is needed, certain goods, such as consumable merchandise, must be stored in a strictly controlled atmosphere and this can be quite easily achieved within an air supported building. In France a fruit grower has managed to keep fruit fresh for as long as 9 months in air supported refrigerated warehouses; this is 3 months longer than he was able to in a more conventional building, and at less than half the capital cost. The ideal atmosphere for the storage of fruit should consist of 92 per cent nitrogen, 3 per cent oxygen and 5 per cent carbon dioxide at a temperature of 3° C. These warehouses were manufactured with nylon fabric, coated on the inside with neoprene and on the outside with hypalon; the latter is expected to give the building a life of more than 15 years. Polyurethane foam stuck to the inside of the membrane with adhesive provides thermal insulation. The inflation equipment supplies a constant artificial atmosphere and a refrigeration plant inside maintains the required temperature. This same warehouse could also be used as an accelerated ripening chamber for fruit, by introducing a highly oxygenated atmosphere and maintaining a constant temperature of around 27° C. Although the portability of these buildings is rather impaired by the insulating foam, this is not critical in this particular application, for which the quick erection and cheapness of pneumatic construction are undoubtedly the main attraction. This application demonstrates the suitability of air supported warehouses, not only for everyday storage, but also for more exacting purposes, where strict environmental control is essential. Judging by the extent of industrial interest, the instant pneumatic warehouse is here to stay.

Plate 92

Manufacturing Industries

Just as instant variation is often needed in warehousing space, so rapid production expansions are sometimes needed. In manufacturing industries productional flexibility can be one of the main factors that secures commercial success. Here again conventional building methods do not allow these rapid expansions; temporary pneumatic building extensions allow an instant increase in production. If this increase is of limited duration then the pneumatic can be dismantled. Otherwise the pneumatic provides temporary enclosure, whilst a permanent building is being constructed.

For many manufacturing processes rigid environmental control is essential, and here the basic concept of air supported construction, the application of environmental energy to achieve structural

Plate 93 Robert Wilson and Sons use a Gourock air supported shelter for a meat processing unit

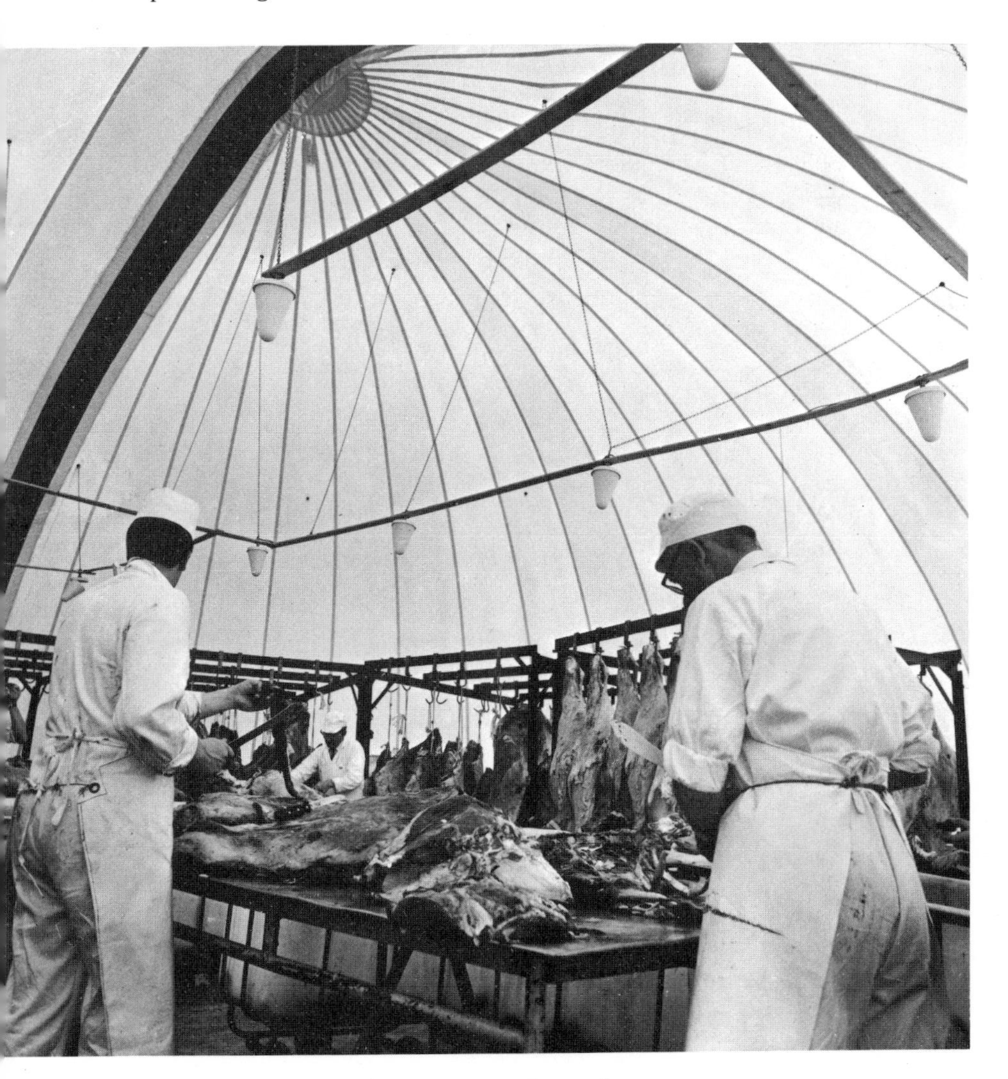

Plate 94 Interior view of meat processing unit

stability, is emphatically demonstrated. Miniaturised electronic industries demand dust-free atmospheres, whilst the preparation of consumable produce requires high standards of hygiene; these are but two of the many manufacturing processes that imply the provision of a manufactured environment, controllable within precise and definite limits, and hence the suitability of pneumatic construction.

One such example is an air supported structure housing a meat processing unit for the food manufacturer, Robert Wilson and Sons, at Kilwinning in Scotland. Besides the low capital cost, two other factors greatly influenced their decision to erect an air supported building; firstly, because of the internal pressurisation, no dust or foreign bodies can enter the building, provided that the air supply is adequately filtered, and secondly the p.v.c. coated nylon membrane can be easily washed, thus allowing acceptable standards of cleanliness. Their satisfaction was clearly portrayed when they ordered an additional central section to double the size of the building. This is indisputably an ingenious solution that provides, for the industrialist, an instant and controllable building environment. Such sophisticated solutions are few and far between.

Plate 93

Plate 94

Industrial applications as yet very rarely exploit the inherent environmental implications of pneumatics. In many cases pneumatics are utilised, not because they afford instant structuring opportunities, but solely because of economic considerations which rule out more conventional construction methods; hence the many diversified industrial uses of pneumatics, whether for storage, manufacture or even maintenance.

Plate 95

Building Construction Industry

In another industrial field, that of building construction, pneumatic enclosures are of major significance. Winter building usually involves extra cost, which can be attributed to loss of time through storms, inefficient labour due to extreme weather conditions, protection of work from the weather and the heating-up of materials to workable temperatures. If the extra money spent on these items was used to provide a completely enclosed construction site most, if not all of this money, could be recovered. With such provision, men would be able to work without interruption, in comfortable conditions, not usually achieved even in summer-time. If such an enclosure can be used repeatedly, the saving in man-hours will almost certainly convert the initial outlay on it into a sound investment. Because of its cheapness and spanning potential, pneumatic construction provides a solution appropriate to these applications and indeed it has already been used with great economic success, whether it be for the erection of radar tracking equipment, the construction of civil engineering projects, or normal building construction.

Plate 96

Plate 97

Plate 95 A Barracuda air supported shelter used for vehicle maintenance

Plate 96 A Barracuda air supported structure allows building construction to continue in all kinds of weather

Plate 97 Construction shelter, fabricated by Birdair, for the radar antennae of the earth tracking station at Andover, Maine

Agricultural and Horticultural Industry

The population explosion is causing urban developments to encroach on present natural agricultural and horticultural regions, which slowly but surely are being severely depleted. Artificial irrigation methods produce fertile land where would otherwise be hot arid deserts. At the other extreme, the cold deserts of polar regions could only be utilised if the environment was artificially controlled within vast enclosures.

The large span potential and the low cost of air supported construction could well make huge greenhouse enclosures a feasible proposition to increase the world's food resources, and to achieve it in otherwise non-utilised regions. During the early sixties investigations were made into the use of air supported greenhouses in the U.S.A., the Netherlands, Germany and Britain. These investigations, admittedly on small span structures, revealed that transparent air supported membranes provided a very satisfactory form of crop protection for the horticultural industry.

Goodyear have recently constructed a giant plastic greenhouse in Ohio, U.S.A., which could well be the forerunner of vast horticultural enclosures for polar regions. The greenhouse, constructed from transparent p.v.c. film 0·3 mm thick and constrained by a galvanised steel cable network is 130 m long, 30 m wide and 6 m high. The main cables, 19 mm in diameter run lengthways and are tied to the ground at 12 m centres. Lighter cables, 4 mm in diameter, run transversely at 1·5 m centres. The internal ties consist of hollow drain pipes which are connected to an underground drainage system. Goodyear is confident that these structures, including the membrane, cables, fans, doors and erection labour will cost about £4 per square metre of area covered regardless of size.

Plate 98

Plate 99

Plate 98 A giant transparent p.v.c. greenhouse restrained by a network of cables

Plate 99 Interior view of greenhouse showing internal tie drainpipes

Plate 100 The world's first air supported office arrives on site

Office Buildings

In the same way that industrial requirements fluctuate, so too do the demands on office accommodation for commercial concerns, but since the environmental requirements for office work are much more exacting than those for the majority of industrial applications, the use of pneumatic buildings as offices has been far from widespread. In fact it was only in January 1970 that the world's first pneumatic office arose out of the snow at Hemel Hempstead, England. This office building, designed by the architects, Foster Associates, for Computer Technology Limited, one of the fastest growing computer manufacturers in Britain, extends the boundaries of pneumatics far beyond mere industrial and sporting enclosures on the one hand and exhibition fantasies on the other. Never have the architectural implications of pneumatics been so conclusively demonstrated; here instant structure is blown up to be stabilised by the application of environmental energy.

Plate 100

Plate 101

Plate 102

One of Computer Technology's greatest problems has been finding building accommodation to keep pace with their rapid growth. For the initial phase of development the architects had luxuriously converted a canning factory. The next phase was originally to expand into newly built accommodation, but official procedures delayed the building programme by 5 months. This caused an acute overspill problem and consequently the architects looked for a temporary solution. Comparisons were made on established temporary building forms, but an air supported structure proved to be by far the most economical answer, less than a third the cost of the cheapest alterna-

Plate 101 Inflation begins

Plate 102 Computer Technology Limited work well into the evening in their new pneumatic office

tive. The air supported structure manufactured from translucent p.v.c. coated nylon, by Polydrom of Sweden, to the architects' specification, is a semi-cylindrical form 60 m long by 12 m wide. Its base has been designed as a future car park, the asphalt surface being covered with polythene, underfelt and carpet. As far as the internal environment was concerned, the architects were seeking standards of sophistication that had very rarely been previously achieved within a pneumatic structure. Heating, ventilating, acoustic and lighting problems had to be resolved, not to mention the fact that local building and fire authorities had to be satisfied.

The functions of heating, ventilating and inflation were combined by a multi-purpose system of two 150 kW oil fired fan heaters, which supply 2·80 m^3 of air per second at an input temperature of 50° C. and this counteracted the cold radiation from the thin membrane to provide an internal air temperature of about 30° C. giving an effective mean radiant temperature of about 20° C.

Plate 103 Section through air supported office

Plate 104 Interior view of office

Apart from the fact that temperature gradients were evident because of poor distribution of the air supply within the building, the performance of this system was found to be quite acceptable. Unfortunately the same cannot be said about the summer environment, when the solar heat gains proved to be very uncomfortable. Despite a sprinkler cooling system which was fitted to minimise these effects and which reduced the ambient temperature by about 5–10 deg. C., internal temperatures as high as 32° C. were often recorded during the summer months.

Lighting and the fire requirements for unimpeded escape in case of collapse were resolved together by means of a double row of lamp standards carrying fluorescent tubes. The light was projected on to the membrane surface and then reflected towards the working area by the white membrane.

Plate 103

Acoustic problems were minimised by reserving the central sound focal area for circulation and by keeping the furniture away from the perimeter; this latter precaution also further reduced the cold radiation effects from the membrane. The carpet and Bürolandschaft type landscaping also helped to improve the acoustics, which on the whole were not obtrusive, although round the perimeter echoes and fan noise were prevalent. However, one acoustic problem proved to be disturbing, that was the noise of the rain thundering against the thin membrane, but although this annoyed the occupants at first they soon got used to it. Surprisingly enough it had one slight advantage, for it gave the occupants an awareness of the external environment in an otherwise completely internal environment.

Plate 104

To allow flexibility of the office planning, the power and telephone take off points were from a perimeter distribution system. The lavish furnishings that were used within this air supported building are undoubtedly a most surprising contrast to the rather stark interiors prevalent in previous pneumatic applications. Without the foresight of the clients, who were willing to explore the unknown implications of a pneumatic office building along with the architects, this building could never have been achieved. It was built to a very strict budget which accounts for some of its slight shortcomings, but the architects' research revealed that these could be easily overcome for very little additional cost. This is surely the prototype for the new instant architecture of the future.

Radomes for the Satellite Communications System

One of the greatest social progresses in recent years has been the rapid technological improvement in communications. With the aid of orbital satellites, such as 'Telstar', a global communications system has been established. The huge radar antennae, associated with this system, require protection from the often alien climate. Rigid radomes of various types have been used in many cases, but their performances have not bettered the standards set by air supported radome structures. To satisfy rigorous design requirements very sophisticated forms of pneumatic construction have been developed. These radomes amply illustrate the technological potential of pneumatic construction, both in terms of refinement of design and in terms of spanning magnitudes.

The largest radome, 64 m in diameter and 49 m high is at Andover, Maine, in the U.S.A. and this simple form boldly defies its rugged environment. A similar radome 49 m in diameter at Raisting, near Munich in Germany, stands serene and proud above the Bavarian

Plate 105

Plate 105 The largest air supported radome at Andover, Maine

plain, its pure white form presenting a perfect foil to the rich green vegetation of the plain. Another radome has been installed at Pleumeur Bodou in Brittany and this survived a 160 k.p.h. wind, a few years ago, which demolished several conventional buildings on the same site.

All these radomes, along with another erected in Nova Scotia, were fabricated by Birdair using a specially developed hypalon coated dacron fabric, which offers very little obstruction to radar transmission. Multiple fans with variable control provide a safe inflation system, and automatically operated air locks afford easy, yet refined, means of access to the building interior. To ensure maximum uniformity of electrical characteristics the radomes were made as one-piece units and these weighed nearly 30 tons. This produced problems for handling and erection. For transportation the membrane was folded along the gore panels into strips 2·5 m wide by 60 m long which were then folded into package of size 2·5 × 12 × 3 m. Erection of the radome before installation of the antennae is considerably easier, the membrane being lifted from its container and spread over the base with a huge crane. This allows attachment of the membrane to the base and subsequent inflation.

However, Krupps manufactured and erected a 39 m diameter

Plate 106 Erection of air supported radome manufactured by Krupp, at Bochum, Germany

radome over an already installed antennae at Bochum, Germany. The membrane was folded in a predetermined manner and lifted by crane on to an auxiliary structure on top of the antennae. The crane then unfolded the membrane until it covered the whole antennae, enabling it to be attached to the anchorage points and inflated. This erection procedure was completely accomplished in $9\frac{1}{2}$ hours.

Plate 106

The standards of sophistication achieved in these large radomes could well be taken as suitable criteria for building regulations, and their pure form will surely be remembered in the future, as a significant contribution to architecture's heritage.

SOCIAL APPLICATIONS

In past eras, social participation has centred around large, permanent and monumental structures, such as the temples and forums of early civilisations, the cathedrals of the middles ages and the civic centres of more modern times. From these centres of social activity, urban communities were generated. The use of pneumatic construction could enable these facilities to change or move with the fluctuations in social patterns. Small, medium or large sized structures, added to 'ad infinitum' could be dismantled and moved to new centres for social participation. Here planners have at their disposal a mobile generator of social activity.

Sport Enclosures

These implications for human, social involvement are already apparent to some extent in the use of pneumatics by many sporting circles. In more northerly communities, such as those in N. America, Scandinavia, Russia and Siberia, severe winter climates greatly restrict sporting activities. In Britain the recent cries have been for all-weather athletic facilities, but economics have only allowed this in a few cases, and these have generally been conversions of derelict military hangars. With the birth of pneumatic building construction, indoor accommodation is now within financial capacity of many sporting institutions.

Until recently garden swimming pools were considered merely as a summer pleasure, but now transparent air supported domes can be cheaply bought that allow this recreation to extend through the winter months. Similarly at local club level and throughout educational institutions air supported enclosures are being widely utilised, permitting such sporting activities as swimming, gymnastics, tennis, athletics, football and cricket, to name but a few, to continue all the year round unhindered by adverse weather conditions. Vast stadii could be easily and economically enclosed; indeed Arthur Quarmby in association with David Powell has proposed the enclosure of Wembley Stadium with an immense transparent air supported membrane. This is an alternative proposal to the helium supported membrane discussed previously. Although no large sporting stadii

Plate 107

Plate 108

Plate 109

Plate 107 Air supported shelter manufactured by Stromeyer, enclosing a swimming pool on the French Riviera at Cannes

Plate 108 Interior view of Cannes swimming pool

Plate 109 Proposals for the enclosure of Wembley Stadium football pitch with a huge air supported structure by architect Arthur Quarmby and plastics consultant David Powell

have yet been enclosed with a pneumatic membrane, the large radomes, and the U.S.A. Pavilion at EXPO '70, have demonstrated that these large spans are not beyond present-day feasibility.

Plate 110

Social Centres

Of great social significance are Manfred Schiedhelm's proposals for the civic centre of Sprendlingen, in Germany, which point at the true potential of pneumatic construction for the enclosure of communal activities. The site for this project is in a public park near the old town centre of Sprendlingen.

Plate 111

'The project is based on the principle of a pneumatic construction, which has been chosen for economic, constructive and aesthetic reasons. The construction consists of a transparent or opaque or sun-protected membrane, which is stabilised in its

Plate 110 Proposed translucent enclosure for tropical birds at Blackpool Zoo, two views. The 62 × 26 m interior solves at low cost the problem of housing creatures requiring a controlled, tropical environment. A 2·5 m high in situ concrete wall acts as a base for landscaping. Architects: Building Design Partnership, with Arthur Hamilton, Borough Engineer, Blackpool

form by a slight over-pressure produced by air conditioning. Large spans can be covered without great effort or costs. Using a contemporary technique a new spatial sensation is created. There are daylight conditions during the day inside the dome. At night lighting projections can be done from inside or outside.

Plate 112

The whole process of building inside the dome can be done during the whole year regardless of weather influences. If the building no longer suits the needs of the community, modifications and extensions can be achieved by adding other spheres using a zip-fastener system. The main aluminium plated steel construction inside the dome can be easily demounted and re-used in the future.'[1]

Accommodation includes, a multi-purpose hall, foyer, restaurant, kitchen, bowling, library, conference rooms and rest rooms, all con-

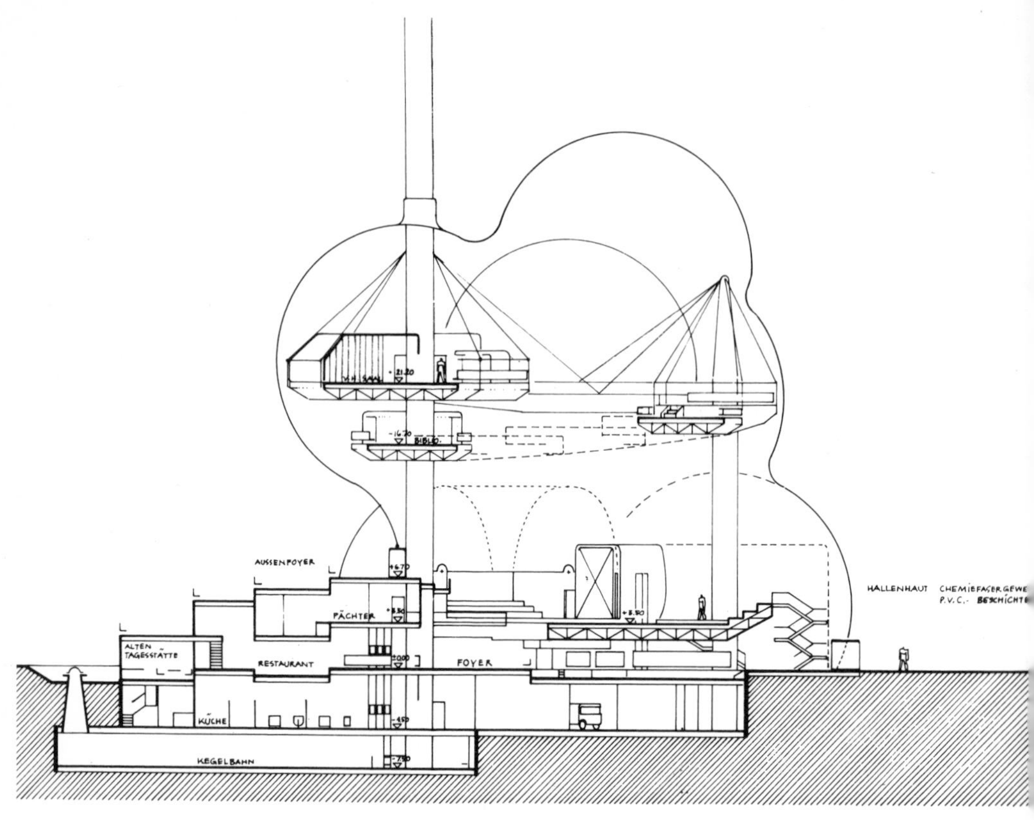

Below: various degrees of usage

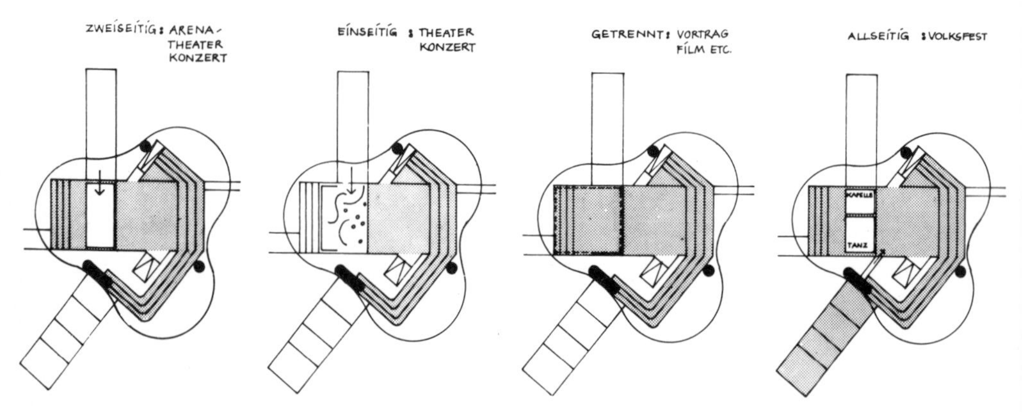

nected by three vertical circulation pylons, one of which terminates in an observation platform offering panoramic views of the surrounding areas. The flexibility of this pneumatic form is adaptable to the rapidly changing requirements of social urban life. Communal structures, such as this, attract and stimulate social intercourse in both urban and rural life, leading to better human understanding.

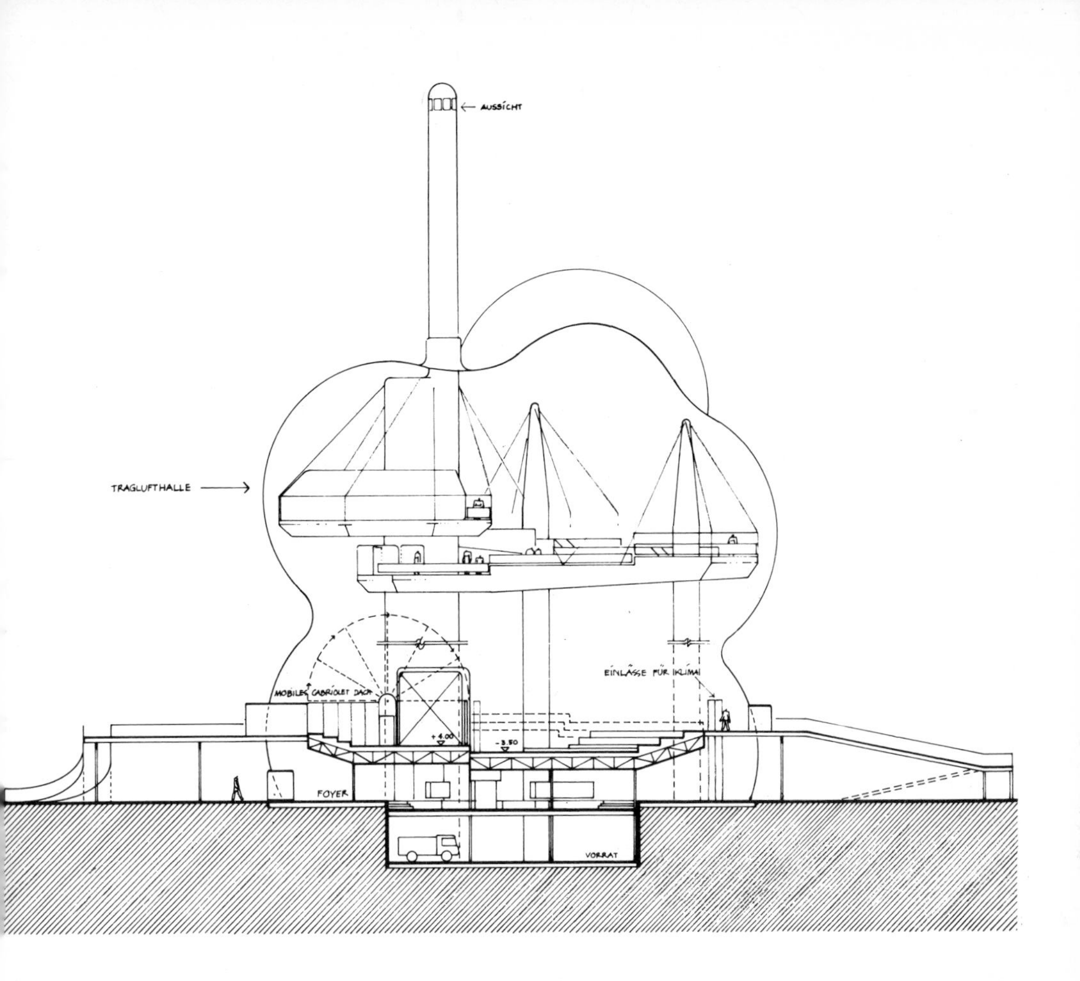

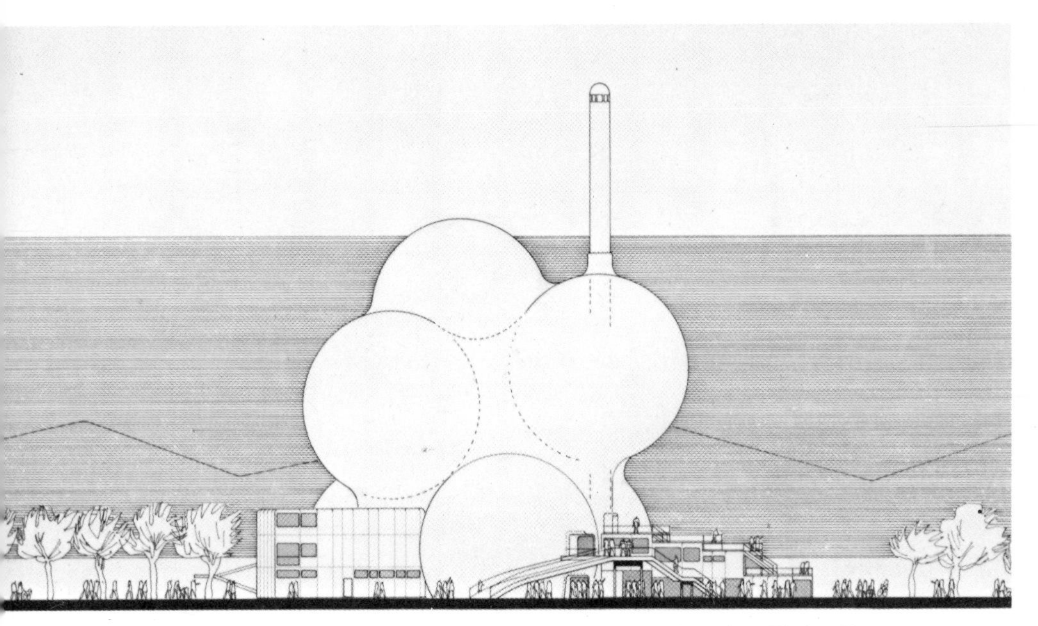

Plates 111, 112 Manfred Schiedhelm's proposals for the Civic Centre of Sprendlingen, Germany; plans, sections and elevation

Plate 113 Krupp's air supported exhibition pavilion at Hanover

EXHIBITION ARCHITECTURE

Despite the very wide and diverse applications for pneumatic building construction, significant architectural achievements have in the main been confined to one application, that is exhibition structures. Unfortunately pneumatic designers have tended to over involve themselves with the purely structural problems of enclosure, and if this tendency continues there is a great danger that the full architectural potential of pneumatics will be lost. It is therefore both refreshing and reassuring to witness designers abandoning their inhibitions for many exhibition structures, but this cannot always be the case.

Travelling Exhibition Pavilions

Many commercial enterprises promote their products with travelling exhibitions that require some form of enclosure whose design must be extremely portable, yet is often restricted by financial considerations. It is difficult to assess the contribution that these enclosures can make to the commercial impact of the exhibits, but as the great exhibition structures of the past have demonstrated this contribution can indeed be significant. Generally it is only the larger enterprises that are sufficiently informed to realise the value and extent of this contribution. Small enterprises are well satisfied with standard production air supported structures which provide a portable enclosure well within their pocket. Larger concerns such as Krupps can afford to be more ambitious with their exhibition structures. In 1966

Plate 113

Plate 114 Air inflated exhibition pavilion designed by Joseph Eldredge for the United States Atomic Energy Commission

Krupp constructed an exhibition pavilion on the Hanover Fair Ground. Here an air supported exhibition hall, 106 m long and 35 m wide and 17·5 m high, was combined with more conventional building forms which housed cinemas, conference rooms and refreshment areas.

The travelling exhibition pavilions for the United States Atomic Energy Commission are even more impressive. The architectural merit of the large one designed by Victor Lundy has already been discussed at length; its outstanding success undoubtedly prompted the client to order another pneumatic pavilion. This smaller structure, designed by Joseph Eldredge and again fabricated by Birdair, resembles a huge inflated pillow that is supported by slender columns. These two pavilions, as well as informing many people about the peaceful uses of atomic energy, have offered these same people a dramatic introduction to this new architecture of pneumatics.

Plate 114

World Expositions

All major international expositions seem to suffer from one major architectural illogicality. Why for exhibition pavilions, which are used for as little as six months, must designers use such permanent and non-reusable materials as reinforced concrete? Many exhibition structures take longer to erect than the duration of the exhibitions, and sometimes take even longer to dismantle. This is absurd, unless the designer can honestly claim to be demonstrating an entirely new

Plate 115 West Germany's Pavilion at EXPO '67 in Montreal

Plate 116 Voguish exploitation of pneumatics on the British Pavilion

constructional technique. For this reason the increased use of temporary and reusable lightweight constructions, such as pneumatics, for exhibition enclosures is to be greatly applauded.

At EXPO '67 in Montreal, pneumatics were conspicuous by their absence, but lightweight structural enthusiasts were by no means disappointed. Frei Otto, the great pneumatic visionary, suspended a beautiful tensile structure over the West German exhibits and this captured everybody's imagination. *Plate 115*

EXPO '70—EXPO PNEU

In contrast to Montreal, EXPO '70, at Osaka in Japan, sprouted pneumatic structures, from the sublime to the ridiculous.

From time to time architects, particularly architectural students, have structural crazes. During the early sixties it was geodesic domes and Buckminster-Fuller was hero-worshipped by students the world over. His dome for the United States Pavilion at the Montreal EXPO heralded the eclipse of the geodesic era. EXPO '70 coincided with the peak of the pneumatic craze, and surely no structural fashion has been so manifest in a single event. Pneumatics, as the new craze, was nowhere more evident than in the feeble inflated Union Jack that hung over the restrained but impeccably detailed British pavilion. However, despite this and many similar abuses, three pneumatic pavilions achieved architectural greatness, one for its subtle beauty, one for its visual prominence, and the other for its technical sophistication. *Plate 116*

United States Pavilion, EXPO '70

The design team for the United States Pavilion, of Davies, Brody, Chermayeff, Geismar and de Harak, from the very outset were intent on erecting a pneumatic pavilion. These designers spent much of their time working in the dark, with no clue as to what the available funds for the pavilion might be. Although their earlier designs were rejected on economic grounds, they remain valuable exercises, demonstrating the possible scope of pneumatics. The first design was for a vast, dual walled air inflated structure, 84 m across that was geometrically between a sphere and a cube. This structure, as well as providing an efficient means of enclosure, would have served as an enormous projection screen in the creation of a sophisticated audio-visual environment. Visitors would have entered by means of a spiral ramp, inside which live theatrical performances would have been in progress. Hence, via a sequence of ramps and platforms they would have ascended to the summit of the exhibition pavilion, viewing on their way a continuous film programme depicting the U.S.A. way of life. On reaching the top they would have then passed through enclosed spaces, in which the main exhibits were to be displayed. From these spaces they would have emerged onto a vast central platform from where they could have witnessed visual representations of space exploration, projected on the roof of the structure. The exodus *Plate 117*

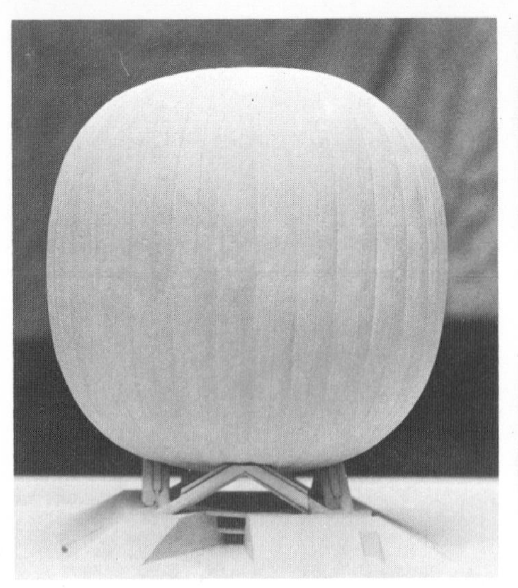

Plate 117 Model of first scheme for U.S.A. Pavilion
Plate 118 Model of second scheme for U.S.A. Pavilion

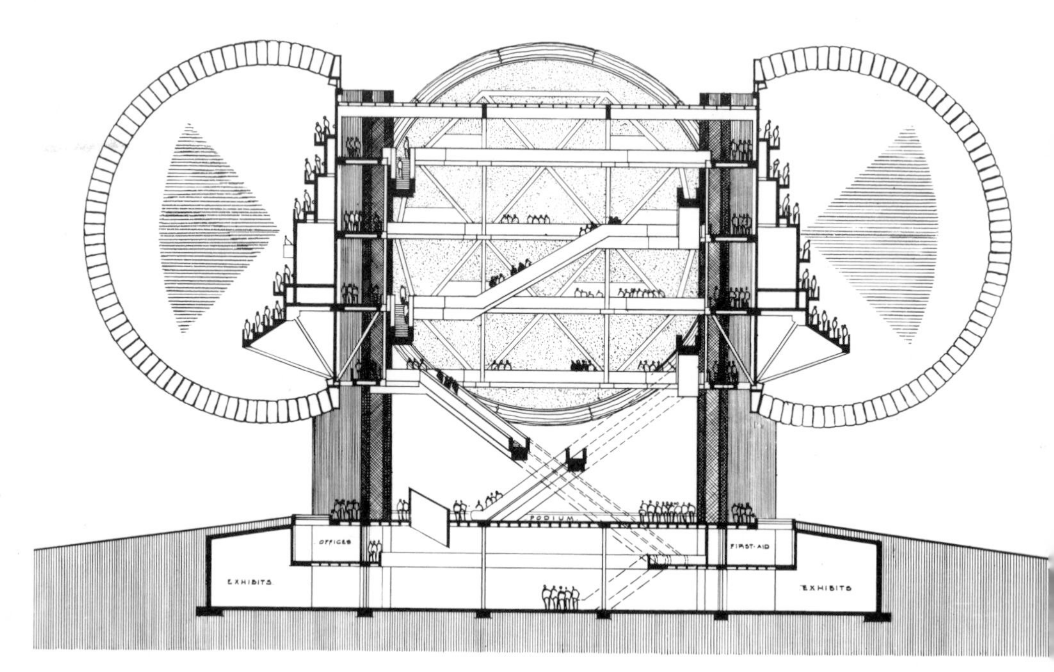

Plate 119 Section through second scheme for U.S.A. Pavilion

would have been by way of a descending spiral from which films projected on the bottom surface of the structure could have been viewed. If this design had been realised there would have been great possibilities for the production of pneumatic sounds, echoing between the dual membranes.

Unfortunately Congress felt that for once, the U.S. economy could not afford this structural extravaganza of over $9 million. Consequently, the designers put forward a new proposal incorporating four pneumatic bubble theatres, in which a similar film programme was presented. Although certainly not as dramatic as the first scheme, much of the environmental impact would have been retained. But even this did not come within the U.S. budget. The original cost allowance of $16 million was finally trimmed to $10 million, $6 million of this allocated to administration, leaving $4 million to cover the cost of the pavilion, its exhibits and transportation to Japan. This was indeed a challenge, considering that 3 years earlier Buckminster-Fuller's geodesic dome, without its exhibits cost $5 million. It is probable that if either of these two ambitious schemes had been adopted, technical problems would have been encountered; no doubt the advisory panel for the pavilion breathed a huge sigh of relief.

Plate 118

Plate 119

The design eventually constructed looked as if the original scheme had suffered a huge puncture. It consisted of a vast oval, cable restrained, air supported roof, 142 m long and 83·5 m wide, which was of such low profile that it was barely visible from outside. Despite Japanese suggestions that this profile reflected the state of American political morale in Asia, it was designed to resist 200 k.p.h. typhoons

Plate 120

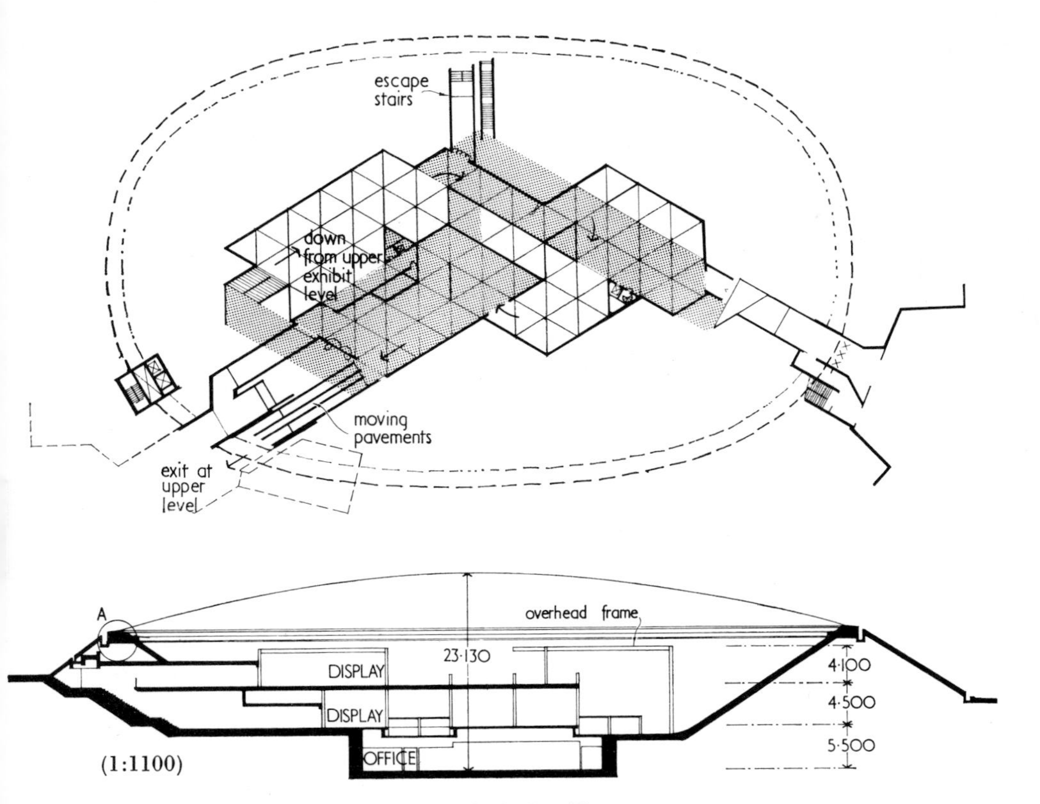

Plate 120 Plan and section of U.S.A. Pavilion

Plate 121 Aerial view of U.S.A. Pavilion

Plate 122 Roof details of U.S.A. Pavilion

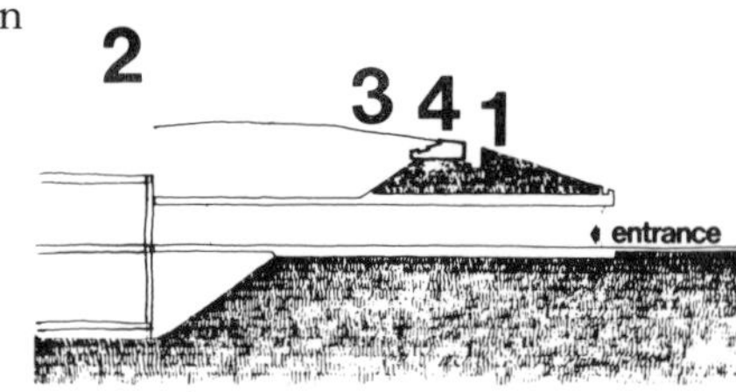

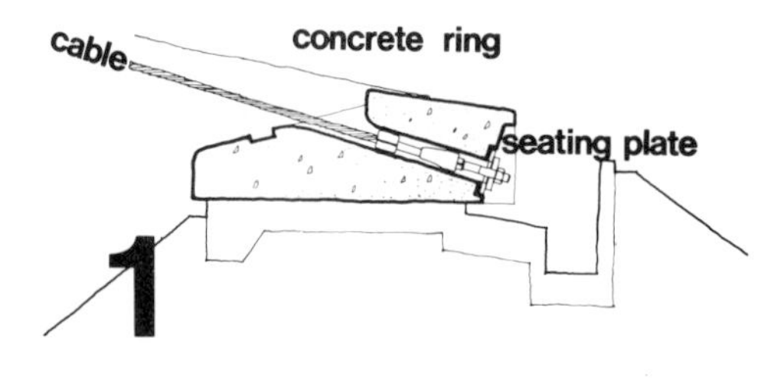

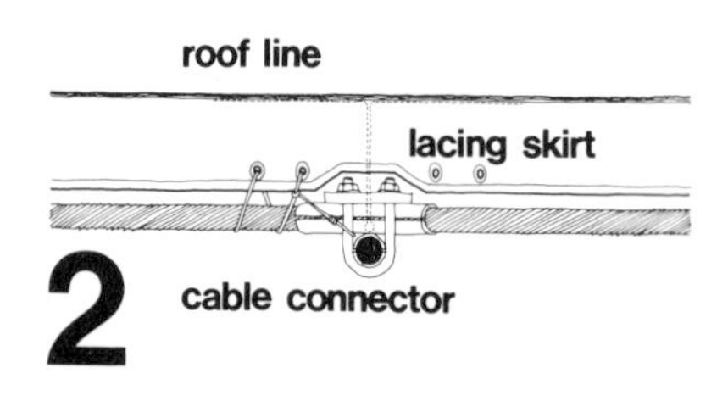

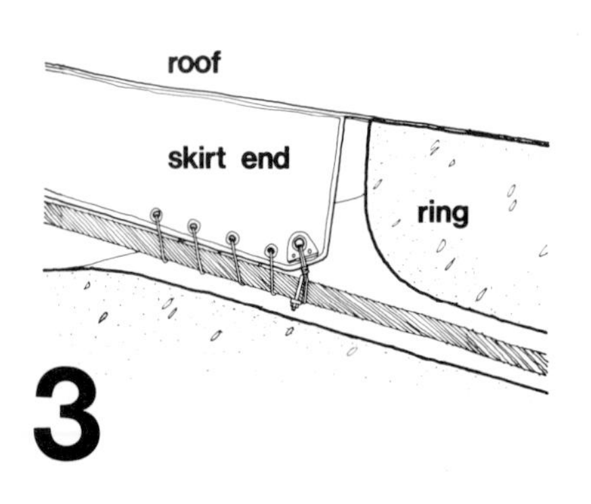

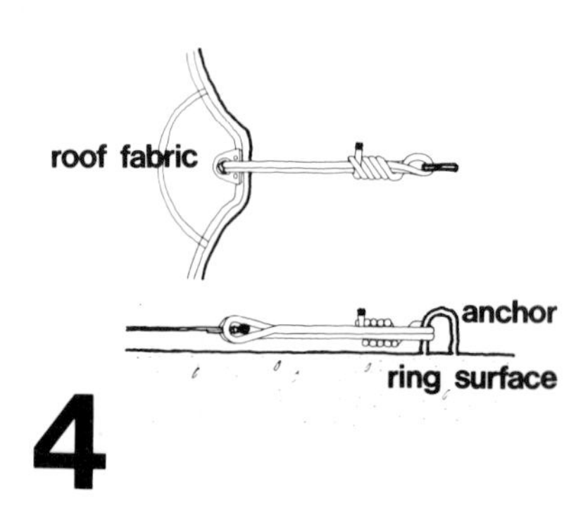

Plate 123 The roof of the U.S.A. Pavilion barely visible from ground level

and earthquakes, and undoubtedly was the most economical structural device for spanning this huge area; the cost of the pavilion was \$2·9 million to cover nearly 10,000 m^2.

The roof membrane consisted of a vinyl coated glass fibre fabric, and it was restrained by a diamond grid cable network, the cables of which ranged from 38 mm to 56 mm diameter. These were anchored in a concrete perimeter ring beam which was cast on top of an earth mound, constructed by excavating the soil from inside the site and piling it up around the periphery; the structure thus took up a sunken posture. The total weight of the roof structure, including the cable network, was 1·2 kg/m^2 of area covered. In contrast Buckminster-Fuller's geodesic at Montreal was a hundred times heavier, and this is surely sufficient to question the advantages of geodesics for large span enclosures. This lightweight roof structure was stabilised, even during typhoons, by an internal pressurisation of only 20 mm of water pressure. This pressure level was maintained by an inflation system of four blowers, each of capacity 3·775 m^3/sec. Two additional standby blowers, of similar capacity, were available for emergencies, and the pavilion also had an emergency generator in case of power failure. These large air volumes were not necessary to maintain structural stability, but to provide sufficient ventilation, and these were ducted to distribution points around the building's perimeter.

Plate 121

Plate 122

Three radial caverns, incorporating revolving doors as airlocks, drew the visitor into the subterranean depths of the pavilion. Out-

Plate 123

Plate 124 U.S.A. Pavilion glows at night

Plate 125 Interior views of the U.S.A. Pavilion

Plate 126 Birdair's air supported structure for the United States Travel Association Pavilion

side the queues became a regular feature not only because of its popularity, but also because of the slow passage of visitors through the revolving doors. The only note of discord, in this otherwise subtle yet impressive interior, was the two-storey, steel framed exhibition structure which pressed against the membrane roof, conflicting against the beautiful space created by the hollowed arena with its reflecting plastic foil rim, and the delicate translucent roof. This unfortunate skeleton structure was designed as a further protection against roof collapse in the event of inflation equipment failure. This interior brilliance was in complete contrast to the unobtrusive, even invisible, exterior. Beyond the asphalt tiled sloping earthwork bank, the gigantic quilt-like structure barely billowed above the sky-line. This subtle yet daring structure contrasted vividly with the insensitive hard-sell monumentality of the Russian pavilion. It demonstrated not only the spanning potential but also the delicate beauty of air supported construction; an indisputed landmark in the history of pneumatics.

Plate 124
Plate 125

Compared with the official United States pavilion, the pavilion for the United States Travel Service achieved nothing. This air supported spherical structure, which clumsily appealed 'Visit U.S.A.' can have done little to further expand U.S. trading. Despite its architectural limitation, like all Birdair fabricated structures it was competently detailed.

Plate 126

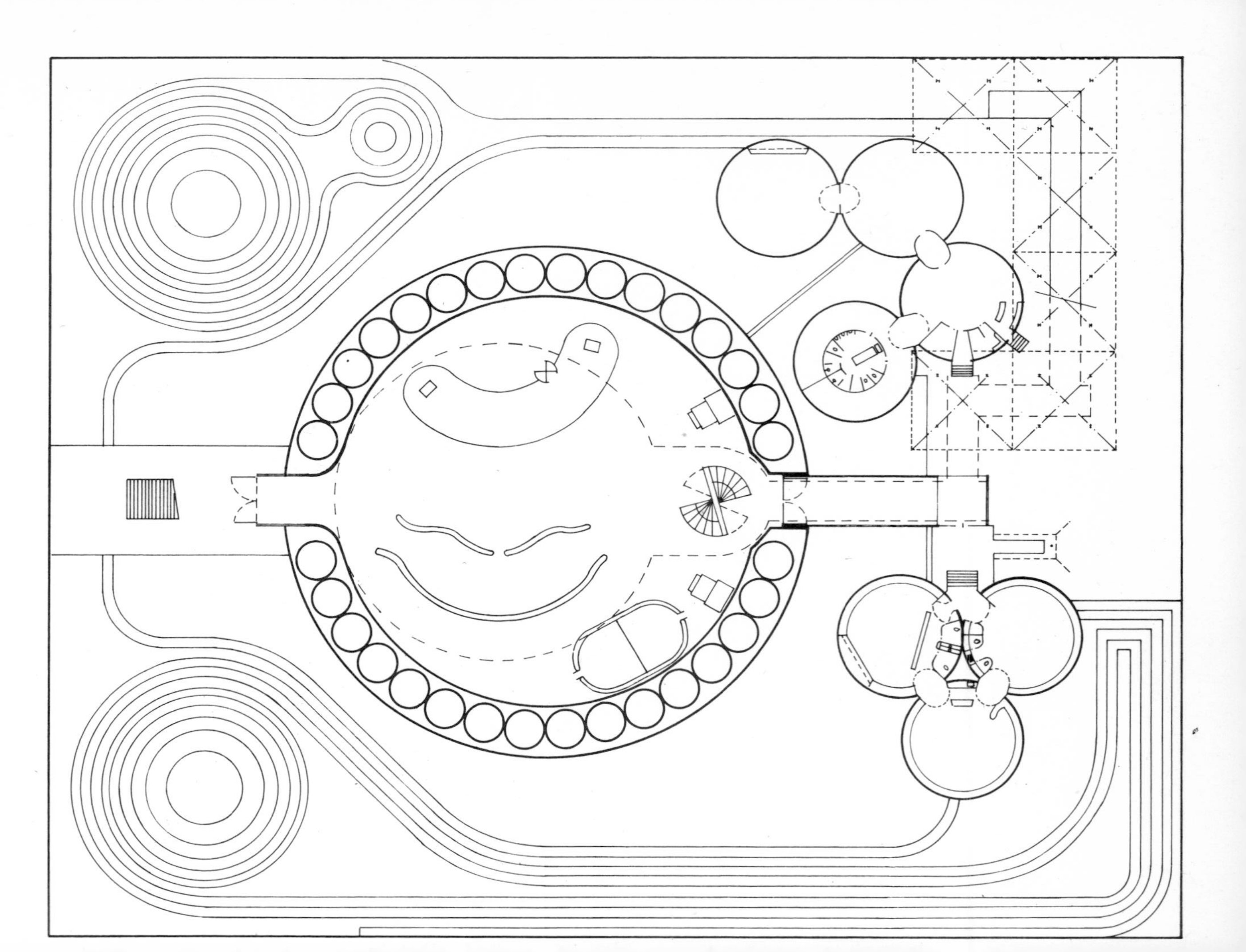

Fig. 90 Plan of Fuji Group Pavilion

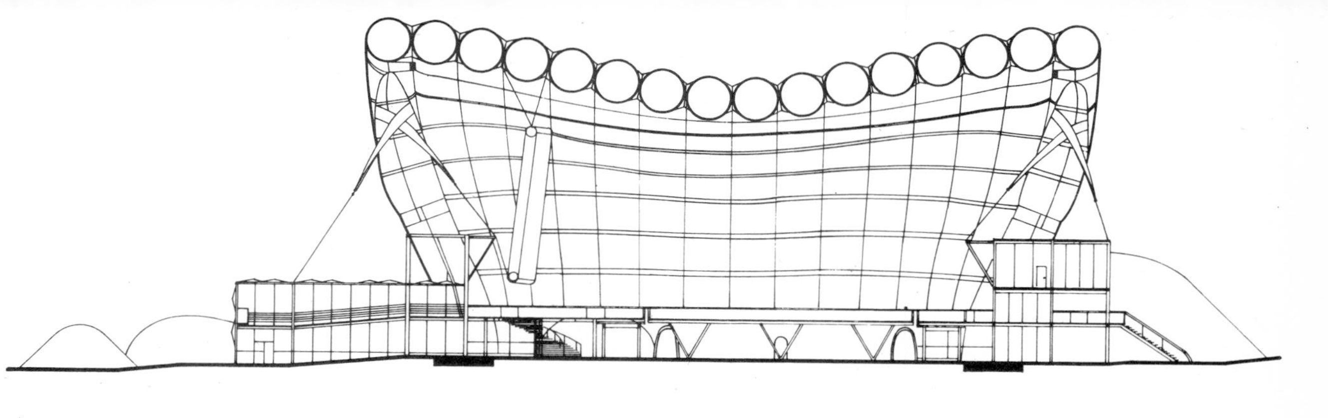

Fig. 91 Section through Fuji Group Pavilion

Plate 127 The dominating Fuji Group Pavilion

Plate 128 An end-on view of the Fuji Group Pavilion

Fuji Group Pavilion, EXPO '70

Fig. 90 Undoubtedly the most spectacular pneumatic structure at Osaka was the Fuji Group Pavilion, designed by architect Yutaka Murata and engineer Mamoru Kawaguchi. Emerging from a circular plan, 50 m *Fig. 91* in diameter, were sixteen air inflated arches, each 78 m long and 4 m in diameter. In the centre these air inflated beams took on a semi- *Plate 127* circular form, but towards the extremities of the structure the ends of the beams were brought closer together, making the apex of the arch higher and causing it to jut forward. A combination of membrane walls and inflated air beams closed off the end of the structure, thus giving it added stability. This obese form strongly resembled a gigantic red and yellow striped tea cosy, and was surrounded by several Binishell type concrete domes.

Plate 128 The large diameter tubes were made out of polyvinyl alcoholic fabric, with an external weatherproof coating of 'hypalon' and an internal p.v.c. lining to reduce air permeation. The fabric tubes were clamped to metal cylinders which were anchored to a concrete base. Separate feeds from a peripheral distribution system supplied compressed air to each tube. The internal pressurisation could be varied between 8000 and 25,000 N/m^2 to suit wind conditions, the higher *Plate 129* level it was claimed would enable the structure to withstand typhoons

Plate 129 Details of the air supply distribution system of the Fuji Group Pavilion

well in excess of 200 k.p.h. Visitors approached this pavilion from the east up a gently sloping ramp, covered by a delicate and brilliantly coloured awning. At the top of this ramp a tiny air inflated reception structure controlled the passage of visitors into the pavilion. Once past this control the visitors were engulfed by the dark internal chasm, where a turntable slowly revolved above a small air supported sausaged-shaped structure, housing the control equipment, and air inflatable restaurant and toilet structures. From the revolving turntable visitors witnessed a highly advanced audio-visual programme. A movie film was projected onto a huge inflatable screen, 20 m wide by 15 m high whilst 28 central slide projectors cast numerous images on the inside of the inflated structural walls. Unfortunately the pavilion was incompatible with this superb audio-visual presentation, which made the static effect of the architectural form redundant and even obstructive. The acoustic shortcomings of the structural form, such as long reverberation times of 6 seconds, prevented the use of many stereophonic effects. So alas, two major technological advances in different fields, structures and audio-visual environments, suffered grossly from an attempt to integrate them and their incompatibilities. Despite this grave shortcoming this pavilion demonstrated clearly both the potential and versatility of pneumatic construction.

Plate 130

Plate 131

Plate 130 Interior views of the Fuji Group Pavilion

Pneumatic Floating Theatre Pavilion, EXPO '70

Plate 132 Murata and Kawaguchi also collaborated on the design for a smaller, but even more sophisticated pneumatic floating theatre pavilion for the Electrical Industries Federation. The whole pavilion was *Fig. 92* mounted on a steel framed base 23 m in diameter and floated on a shallow lake, supported by 48 buoyancy bags. The amount of air in each bag was automatically adjusted to respond to the audience movements above. In this way listing was prevented. During each 20 minute performance the floating structure rotated slowly across the lake.

The theatre was enclosed by a roof membrane consisting of p.v.c. *Plate 133* coated double woven polyester fabric, which was supported by three large air inflated beams, 3 m in diameter and forming arches 23 m in diameter. The internal pressurisation could be varied between 15,000 and 30,000 N/m^2, depending on wind conditions. The acoustically formed ceiling to the auditorium was achieved by attaching a light polyester membrane to the undersides of the arches, and partially vacating the air between this and the roof membrane. The resulting negative pressure differential, 10 mm of water pressure below atmospheric, supported the ceiling and further stabilised the overall structure. This use of a negative rather than a positive pressure differential was certainly an innovation, which demonstrated conclusively that pneumatic construction need not be restricted to simple structural forms.

Plate 131 Erection of the Fuji Group Pavilion

Plate 132 Model of the pneumatic floating theatre pavilion for the Electrical Industries Federation

Plate 133 Pneumatic floating theatre pavilion

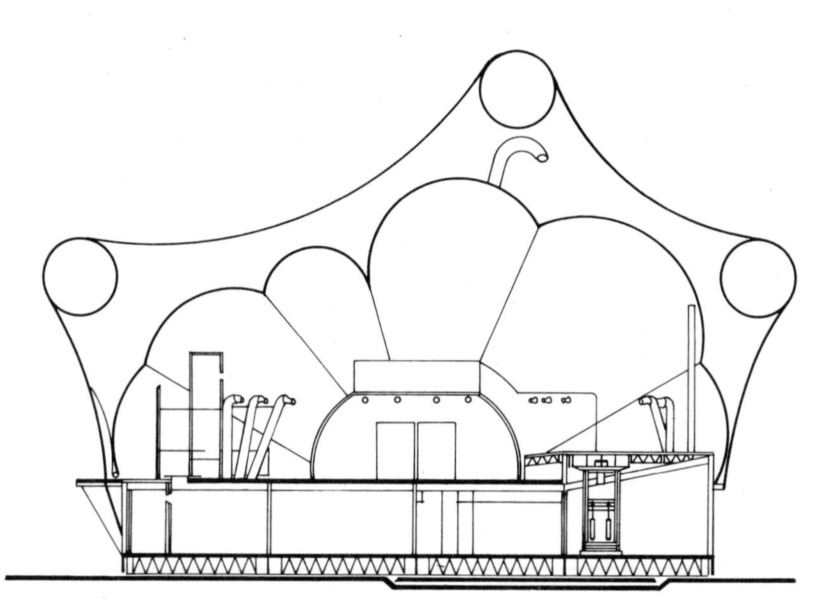

Fig. 92

Plate 134 Air inflated polyester roof covering to Kenzo Tange's immense space frame roof covering EXPO '70's Festival Plaza

Originally the pavilion was designed with some air inflatable seats, but they gave so much trouble during the first operational week, that they were discarded and replaced by ordinary chairs. Despite this failure, this was undoubtedly one of the most sophisticated pneumatic structures that has ever been built, and it is gratifying to note that the Japanese Ministry of Science and Technology has awarded Murata a special medal to commemorate his pioneer work on pneumatic structures.

More Pneumatic Structures, EXPO '70

These three buildings, the United States Pavilion, the Fuji Group Pavilion, and the Floating Theatre Pavilion were without doubt the most impressive pneumatic structures at EXPO '70. Their contribution to pneumatic development is indeed significant; never before had such sophistication and potential been so dramatically expressed in pneumatic construction as here. Set beside these, the abundance of other pneumatic structures of all shapes and sizes at Osaka are of little significance, except by the fact that this abundant exploitation did much to familiarise the many EXPO '70 visitors with a new building construction technique. Many of these are merely carnival structures utilising the latest structural craze.

Originally the French pavilion was to be an air supported construction, but this never materialised. Jean Le Couteur and Denis

Plate 135 Information pavilion at EXPO '70

Sloan, the winners of an open competition for the pavilion, designed a four dome air supported complex, three intersecting to form one enclosure and one detached. Difficulties were encountered finding a suitable material and developing satisfactory jointing methods between contiguous domes. By the time these had been overcome costs had soared and completion could not be guaranteed, so the organisers abandoned air supported construction and invited the Japanese associate architects to take over and use more conventional structural techniques. It is a pity that the air supported concept was not realised

Plate 136 The Ricoh Pavilion EXPO '70

since construction knowledge on jointing methods would have been considerably advanced.

Many other inflatables were evident, but these could hardly claim to be architecture; these were simply exposition hardware. Inflatable roof elements, inflatable umbrella pavilions and inflatable information kiosks were scattered throughout the EXPO' site, all visually prominent, some attractive, others verging on the ridiculous. One, part of the Mitsui Group Pavilion, was billed 'The greatest atmospheric structure in the history of world expositions', but it was so

Plate 134
Plate 135
Plate 136
Plate 137

Plate 137 Air inflatable umbrellas in Expoland

Plate 138 obscene, that it can only be regarded as a huge joke. It is this sort of thing that is detrimental to pneumatic progress. However, despite this and other atrocities, EXPO '70's pneumatic exploitation was so great, that the potential of pneumatics has been more clearly brought to the notice of the whole world; architectural subtleties, spanning opportunities and technological sophistication were all evident at EXPO '70.

Plate 138 'The greatest atmospheric structure in the history of world expositions'

PNEUMATIC ARCHITECTURE OF THE FUTURE

The dynamism of pneumatics coupled with their do-it-yourself experimental potential has fostered a blow-up craze in the field of architectural education which has extended not only into architectural magazines but also into the daily national press. Although this publicity has extended the familiarity of pneumatics, it has in some cases been detrimental to further development, for such attempts to attract publicity very rarely seriously consider the detailed practicalities of pneumatic use. It is therefore not surprising that public opinion has been rather sceptical about pneumatics. However, it appears that this voguish way of looking at the subject is on the wane and with it will go many of the frivolous proposals that have been made for pneumatic application.

Undoubtedly the greatest potential for pneumatics appears to be in the field of instant structuring. The degree of sophistication and efficiency of inflatable survival equipment, such as the life-raft, are both evidence that instant pneumatic homes are by no means an impossibility. From man's continual yearning for change will no doubt emerge a completely mobile home, and indeed pneumatic construction would appear to offer one of the best solutions. Arthur C. Clarke predicted in 1963 that man will return to his nomadic tendencies of the earliest civilisation although on an infinitely more advanced technological level.[2] Clarke's fully mobile home would require not only a self-contained propulsion system but also self-contained power, communications and other services. Although this may not at present be feasible, it will almost certainly not be beyond tomorrow's technology. These predictions of Clarke although undoubtedly utopian in 1963, are now a distinct reality judging by present-day social trends. There is a distinct desire for continuous change, a desire to keep altering one's surroundings. This, as mentioned in the introduction, is observed in two ways, firstly the greater turnover in material belongings, such as houses, cars and furniture, and secondly in the greater movement from one environment to another made possible by better communications. These trends imply either a mobile architecture or an architecture with a much shorter life span than at present, perhaps even a throw-away architecture planned for obsolescence after a specific time. Such short-life buildings must ultimately be realised in some form of instant structure in which on-site erection time is reduced to a minimum. At the moment the inflatable survival life-raft is the ultimate in instant structuring, contained when not in use in a very small volume, yet inflating in seconds to a structure many times its package volume. Undoubtedly this form of construction could be one answer to Clarke's predictions: instant structuring provided by air supported or air inflated forms and mobility provided by some form of air controlled structure similar to the ground effects machine which would free movement from the fixed transportation routes.

Such revolutions are for the future times when people are no longer intent on clinging to traditional living concepts. Unlike more conventional structures, the pneumatic is alive, reacting noticeably to applied loads which attempt to deform it. The pressurisation media contained within its membrane, struggle to generate spherical forms, and to burst through this restraining membrane. Its kinetic character makes it one of the most exciting architectural forms ever conceived and its potential is indeed extensive. In these early stages of development it is essential that we do not abuse pneumatics by using them as novelties, but employ them where they provide an efficient solution to a particular building task. They have a mind of their own and only if we control their design, fabrication and erection with great caution, will failures be avoided and pneumatic potential realised.

REFERENCES

1. M. Schiedhelm, Berlin, October 1967.
2. A. C. Clarke, *Profiles of the Future*.

Pneumatics Design Guide

1 Determine viability of pneumatics for building type

Possible building types for which pneumatics have been proved viable:
- Portable Buildings
- Temporary Buildings
- Instant Buildings
- Low-Cost Buildings
- Large-Span Buildings

2 Establish compatibility of function with pneumatics

Pneumatics already used for following applications:
- Military
- Commercial—Warehousing
  - Manufacturing Industries
  - Building Construction Industries
  - Agricultural and Horticultural Industries
  - Office Accommodation
- Social —Housing
  - Sport
  - Community Centres
- Exhibition Architecture

Pneumatics considered for mobile architecture

3 Determine environmental design criteria

- Thermal Control
  - —by manufacturing artificial environment; consider
    - heating
    - cooling
    - ventilation
    - insulation
  - —by passive control of solar and terrestrial radiation with multi-wall constructions
- Visual Environment
  - Daylight—consideration of translucent and transparent membrane walls
  - Artificial
- Acoustic Control
  - —by plan zoning
  - —by spatial forms
  - —by absorption and reflection of sound

4 Determine pneumatic form and construction type

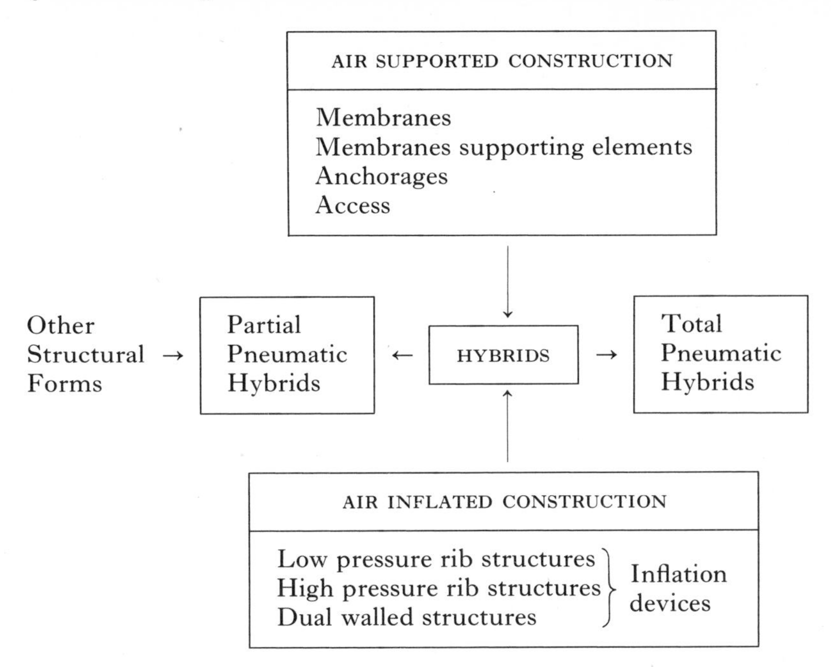

5 Design of pneumatic structure

Choice of suitable pneumatic forms using
- Soap films and bubbles
- Frei Otto's Laws of formation
- Model analysis

Structural design considering
- Loadings

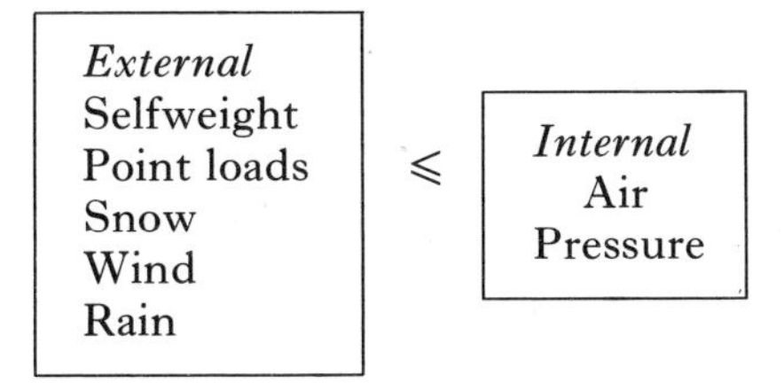

- Membrane stresses—use of cables, cable nets and internal membrane walls
- Boundary conditions
- Anchorages
- Inflation equipment

Choice of suitable safety factors
Choice of suitable materials and equipment

6 Legislation

At present there are no set standards controlling pneumatic design. Interpretation of existing regulations for conventional building design is at discretion of local authorities. Future regulations should consider
- Structural design of membrane and anchorage
- Performance of materials and jointing methods

Design of inflation equipment as regards
- Capacity based on air losses
- Duplication or standby equipment
- Ventilation requirements

Fire precautions as regards
- Ample escape provision
- Material incombustibility

Access provisions

Bibliography

Although in recent years pneumatics have featured prominently in established architectural magazines, there still remain only two major works of reference.

1. Otto, Frei. *Zugbeanspruchte Konstruktionen*, Band 1. Berlin West 1962, Ullstein Fachverlag. This the first book to be published on pneumatics has since been translated into English—*Tensile Structures—Volume 1*. M.I.T. Press, Cambridge, Mass., 1967.
2. *Proceedings of the 1st International Colloquium on Pneumatic Structures*. University of Stuttgart, 1967. This is a record of the first and as yet only major international conference on the subject. Copies can be obtained from Institut für Modellstatik Technische Hochschule, 7 Stuttgart, Postfach 560, Germany.

Technical papers which have been consulted have been acknowledged in the references at the end of each chapter. Besides these the following references have been a source of information.

3. Banham Reyner. *The Architecture of the Well-tempered Environment*, Architectural Press, London, 1969.
4. Boys, C. V. *Soap Bubbles, their formation and the forces that mould them*, Heinemann, London, 1960.
5. Clarke, Arthur C. *Profiles of the Future*, Gollancz, London, 1962.
6. Clarke, Basil. *The History of Airships*, Jenkins, London, 1961.
7. Daugherty, Charles Michael, *City Under the Ice, the story of Camp Century*, Macmillan, New York, 1963.
8. Dent, Roger N. *Pneumatic Structures in Architecture with Special Reference to Arctic and Lunar Applications*, dissertation, University of Liverpool, 1968.
9. Dollfus, C. *Balloons*, Prentice-Hall, London, 1962.
10. Fuchs, Sir V. E. and Hillary, Sir E. *The Crossing of Antarctica*, Cassells, London, 1958.
11. Lanchester, F. W. *Patent 119,339*, London, 1917.
12. Lanchester, F. W. *Span*, Transaction of the Manchester Association of Engineers, Manchester, 1938.
13. *Pneumatic Building Construction*, the Academy of Building and Architecture, U.S.S.R., Moscow, 1963 (Russian text).

Photo Credits

Acknowledgements are due to those who provided photographs, the sources of which are listed below.

Pneumatic Tent Company: 1, 7. Birdair Structures Inc: 2, 11, 12, 13, 17, 40, 41, 88, 97, 105, 126. Barracudaverken: 3, 16, 29, 43, 45, 46, 47, 48, 49, 56, 58, 91, 95, 96. Krupp: 4, 54, 106, 113. Don Cameron Balloons: 5. Goodyear International Corporation: 6, 74, 85, 86, 98, 99. Lea Bridge Industries Ltd: 8. Ministry of Technology, Research and Development Establishment, Cardington; 9, 67, 68, 69, 70. R.F.D.-G.Q. Ltd: 10, 59, 60, 66, 71, 72, 73. ML Aviation Company Ltd: 14, 15, 80, 81. Victor Lundy: 18, 19, 20, 21. British Hovercraft Corporation Ltd: 22. George Cserna: 23, 24. J. M. L. Dent: 25, 26, 27, 28. Stromeyer: 30, 107, 108, 115. John Pye: 31. Cross and Ticher Ltd for Union GmbH, G.D.R.: 32, 33, 36. Francis Thompson Studios, courtesy of British Ropes Ltd: 34. Montgomery, courtesy of Arthur Quarmby: 37, 38, 109. Roger Dent: 39. Gourock Ropework Co Ltd: 35, 42, 50, 55, 93. Du Pont: 44, 62, 63, 92. N. Laing, *Proceedings of the 1st International Colloquium on Pneumatic Structures*, University of Stuttgart, 1967: 51, 52, 53. Robert Wilson and Sons Ltd: 57, 94. Rex Lowden, courtesy of Frankenstein Group Ltd: 61, 64. Bayer Leverkusen: 65. Firestone: 75, 76. Pat Hunt, courtesy of Goods and Chattels Ltd: 77. The Macmillan Company: 78. Transantarctic Expedition, 1958: 79. Garrett Corporation: 82, 89 90. National Aeronautics and Space Administration: 83, 84. Controller, H.M.S.O.: 87. Foster Associates: 100, 103. John Donat, courtesy of Foster Associates: 101. Gus Cordl, do: 102. Tim Street-Porter, do: 104. Manfred Schiedhelm: 111, 112. Joseph Eldredge: 114. Davis, Brody, Chermayeff, Geismar, de Harak Associates, Exhibit Design Team: 117, 118, 119, 120, 121. Arai, courtesy of above: 122, 123, 124, 125. Selwyn Goldsmith, courtesy of *Architects' Journal*: 127, 133. Tim Street-Porter: 116, 128, 129, 130, 134, 135, 136, 137, 138. Taiyo Kogyo Co Ltd: 131. Yukata Murata: 132. Building Design Partnership: 110.

Index